Claire Bessant spent twenty-eigh[...] of International Cat Care, a charity that aims to make the world more cat-friendly in all aspects of our interactions with them, from veterinary care to understanding our pet cats, to finding the best ways to help unowned cats. She has authored many cat books over the years including *The Nine Life Cat, What Cats Want, The Perfect Kitten, The Complete Guide to the Cat, The Ultrafit Older Cat, Cat – the Complete Guide, The Complete Book of the Cat* and *The Haynes Cat Manual.*

CLAIRE BESSANT, The Cat Whisperer

How to Talk to Your Cat

GET INSIDE THE MIND OF YOUR PET

jb

First published in the UK by John Blake
An imprint of Bonnier Books UK
4th Floor, Victoria House,
Bloomsbury Square,
London, WC1B 4DA

Owned by Bonnier Books
Sveavägen 56, Stockholm, Sweden

facebook.com/johnblakebooks
twitter.com/jblakebooks

Paperback: 978-178946-599-0
Ebook: 978-1-78946-656-0

A CIP catalogue of this book is available from the British Library.

Designed by www.envydesign.co.uk
Printed and bound by Clays Ltd, Elcograf S.p.A

1 3 5 7 9 10 8 6 4 2

Copyright © Claire Bessant, 2023

First published in paperback in 2004
This edition published in 2023
This edition has been substantially updated for publication in 2023.

Claire Bessant has asserted her moral right to be identified as the author of this Work
in accordance with the Copyright, Designs and Patents Act 1988.

John Blake is an imprint of Bonnier Books UK
www.bonnierbooks.co.uk

For Leo, Lucy, Mike, Sinead and Sorcha and their
wonderful approach to all the cats we have had over the
years, and of course to Chilli, Mr White and Mello,
who continue to teach us to 'think cat'

Contents

6. What does liking cats say about us and what do cats think of us?

Introduction

When I was first asked if I would like to update my book *How to Talk to Your Cat*, I thought the timing was perfect. The original book was published in 1988 pre-children and pre-becoming chief executive of the charity International Cat Care. Widespread interest in cat behaviour was in its early days and the first thoughts about 'problem behaviours' in cats within our homes (behaviours that are problems for owners but natural for cats, but which usually don't occur in the house, such as soiling) and discussions on their causes and solutions had only just started, and with that an increased interest in why cats do what they do in general. Cat knowledge was behind that of dogs, but there were some pioneers on understanding cat behaviour and interest was growing.

Over the past thirty years or so our knowledge of cat behaviour has improved considerably, more research is taking place and we are continually gaining insights. I have been lucky in that the charity I worked for has a broad base of knowledge of cats in all the situations in which we live and work with them; it is heavily invested in cat health via its veterinary division. It is so important to understand both physical and mental health, which are so closely associated in cats. It also brings a scientific approach to trying to figure out what is best for cats, being sympathetic to their feelings but not over-emotional, as well as being humble and open to understanding ourselves and our needs. We also have to be pragmatic – perfect is not possible, but a balanced way of looking at cats and how what we do can affect them, and how we can best communicate with them, can help us to live together in harmony.

Cat lovers have always said that their cats each have their own unique personalities and can be very individual and different to each other. Over the years I worked for International Cat Care, the charity took a strong lead on understanding cat behaviour and using that to find solutions to improve the care of cats – from population management to living closely with our pets, breeding, veterinary care and any way we interact with them.

I have been lucky to work with some brilliant and famous cat behaviour experts including Vicky Halls (who wrote *Cat Detective* and *Cat Confidential*) and Sarah Ellis (who wrote *The Trainable Cat*), as well as world-renowned specialists in feline health. We explored the cat/person relationship and kept abreast of new findings, aiming to provide pragmatic interpretation of any discoveries for cat owners and people who work with cats. Vicky coined a phrase which is now one of the charity's Cat-Friendly Principles: 'Respect the species, understand the individual', which pulls together our broad understanding of what cats are with what makes each one behave as it does – what its personality is, what it likes and dislikes, what its particular needs are, and how we adapt our interactions to care best for our cats and improve our relationships with them. Throughout this book I will proudly refer to the work of International Cat Care and the people who have contributed so much to our understanding of cats and their welfare. It is an amazing charity which has led on our thinking about cats over the years and continues to do so. Its website, www.icatcare.org, is a goldmine for people wanting to learn about cats.

The title of the book refers to talking to your cat because, being humans, we fall back on vocal communication a great deal due to our own rich language skill and vocabulary. We forget that when we talk to people we don't just listen to their words in order to understand them – we are also taking in body language, facial clues and tone of voice. We are using our knowledge of our own species to interpret what is happening, using empathy and knowledge of what is important to people and what they are likely to do. We can't communicate successfully without that understanding. So when we think about 'talking' to our cats, having as much insight as possible to help interpret the communication between us is vital. Our 'talk' won't just involve words, but actions, and the best reactions to what they are doing, taking into account how their environment will also be influencing them. Communicating with a different species is a little complicated; we need to be detectives and to build up a picture of our cats and how they behave as well as knowledge about the species in general. Knowing how your cat normally behaves and reacts is extremely useful, not just to understanding it but noticing when behaviour changes perhaps because of illness or something that is causing the cat to be unsettled or stressed.

I have recently retired from International Cat Care, the charity which has been a part of my life for over twenty-eight years. By chance one of the speakers at the charity's veterinary congress just before I retired was Dr Dennis Turner, one of the early researchers and writers on cats and their behaviour – his book, with Patrick Bateson, called *The Domestic Cat: The Biology of its Behaviour*, is on the shelves of anyone interested in cat behaviour and has been on my shelves for over thirty years, so it was an honour to meet him. During his lecture he mentioned in passing that 'cats are the most individual of all animals' and that really caught my attention. I asked him to elaborate – did he mean that cats are more individual in their behaviour than other animals? I hadn't really thought about cats in those terms. He answered as the scientist that he is:

In each and every multivariate analysis of cat data in our studies 'individual differences' (individuality) have always turned out to be the most significant effect and we have had to 'isolate' that to determine the influence of other variables. No studies of dog (or other species') behaviour that have been published (that I am aware of) have found or mentioned that.

That all sounds very scientific and statistical (which is necessary in research), but points to cats having strong individual characters which affect how they behave. Researching human behaviour is not easy, and it sounds like researching cat behaviour is even more difficult! But the individualism made absolute sense to me and was even more relevant considering the charity's 'Respect the species, understand the individual' approach – it brought it all together. It also really got me thinking about what that means to us as owners and how we take the information which is now available about cats, their health and their behaviour and apply it to our own cats at home.

The first edition of this book contained a lot of facts and figures, such as the number of muscles required to move the ears or the frequency at which cats hear, which I have mostly omitted in this revision in order to focus on a more psychological approach to our understanding of cats. My aim is to get us all thinking about cats and pick up a 'feel' for their personalities, what we are saying to them (not just words, but via body language and our behaviour), to encourage us to watch how they react to us and consider how we react in response to the cues they give us. I want us to be able to use our senses and be curious about our cats. We will see in some of the research referred to that people have tried to give cats labels according to the type of personality they have. This is necessary when you are trying to navigate your way through all the things cats do and figure out why they do them and what influences a behaviour. But even though we love a label and a box to put people or pets into, it can sometimes be too confining and limit how we interpret what they do and also our ability to see differences in behaviour caused by circumstances or health.

Cats are complex and often don't give much away, but they do let us know quite a lot if we are listening and noticing. I am not an expert researcher, nor do I have any qualifications in animal behaviour, but over the years I have worked with many people who have these talents and I am often the one to ask the stupid or the 'But why?' questions. Over those years I have also lived with a number of cats and watched them, how their behaviour changes over the years and how they react to things around them. Of course, some cats don't like to be watched closely or stared at, so I have tried to watch in a gentle way that doesn't affect what they're doing.

We can try to be clever and analyse every miaow or tiny movement of the ears, but it is important to get an overview so that we can relax and enjoy our cats, taking each insight as a reward into understanding another creature's world, one that is very different to our own but which seems to be able to share ours very successfully if the circumstances are right. The aim is to understand your cats and enjoy them interacting with you – however distant or close that relationship might be. Updating this book and bringing together all the ideas in a hopefully coherent way has made me look at my cats anew and ask the 'But why?' questions of myself.

The approach I am going to take is to get a feel for how cats sense the world, to be able to recognise some basic behaviour and body language, and to understand what affects that behaviour and what forges a cat's personality. This we will do without reference to people to begin with, just to set the foundations and get a feel for the 'essence' of cats. Then we will begin to bring in that complication of the cat/owner relationship – us! How we affect our cats' lives, how we can interpret the communication between our two very different species (as well as we can), and how we can start to look at our own cats very differently – with an open mind and the joy of understanding just some small parts of what the mind and body conversation between us is actually communicating. It is not an instant thing, and the longer we live with our cats, the closer that understanding becomes. And, of course, a new cat may have a very different personality and may require us to build up all of that information and insight again. However, if we are listening and talking in the right way, the process will be a lot faster!

We will also look at what cats need and want, and consider what they might think of us; to try and pick apart the foibles of our human approach and the effect of things such as breeding or taking an anthropomorphic approach, neither of which, on occasion, is particularly cat-friendly. By the end I hope that we will feel confident to pick up cues from our own cat and react in a way that makes our relationship closer and more enjoyable.

While putting this book together I have included my thoughts more than previously, probably because I have had so many years since of thinking about cats every day and developing strong feelings on some areas in particular. Some, to be honest, are probably based on frustrations about our attitudes, which sometimes do not take our cats into account as much as they should. I have included some of this in the book and have used observations of my own cats as examples in the final chapter – these are anecdotes based on three animals. Analysing my current cats' behaviour certainly can't be called proper research or science, but I will use them to illustrate and to question what a behaviour might mean. They are certainly all individual in their approach to life and it's these differences that can help us to learn about all cats. Research has to look at single behaviours and try to isolate what affects them, but of course it is much more complicated in our daily lives with our own cats and needs to be interpreted in a pragmatic and sympathetic way.

I have referred to our cats as 'it' or 'they' rather than 'he' or 'she' – readers will know it is not because I think of cats merely as objects, and anyway, I don't need to use 'he' or 'she' to make people reading this book more empathetic towards cats. I hope it is obvious that I love and respect cats, so I make no apology for this and hope that the essence of the message is more important. I have used 'owner' instead of caregiver or caretaker or guardian perhaps because I am used to doing so, and because most people understand and don't need to be persuaded that living with our cats goes well past 'owning them' into caring for them as best we can and that our cats are usually in charge of us!

When things are common in our lives, we can start to take them for granted – cats are now kept as pets all over the world. What would an alien think of one species living closely with another so unlike it? Perhaps we need to stand back and look from that point of view to see what is logical and what perhaps is rather strange or one-sided – or makes no sense at all! The cat is popular on social media all over the world and what we see may be extremes of cat behaviour, such as fear or deep affection. Does this bring with it expectations of all cats being the same without considering what a cat really is? Have our attitudes to cats changed and how do those attitudes affect our cats and our approach to them?

In around a hundred years cats have gone from being creatures that lived near us and acted as pest controllers to being members of the family. We have expectations of how this remarkably adaptable, yet fundamentally not quite domesticated, animal should live with us. *How to Talk to Your Cat* focuses on what we humans may think is the ultimate communication, through speech, but what happens at that pinnacle is built upon understanding, empathy, patience, receptivity and respect for this fascinating creature.

1.
The 'essence' of cats

WE SEE CATS every day, in our homes, on our streets and on social media. Because they are so familiar to us, perhaps we assume we know what they are by viewing them through some of these filters. Does this mean we take them for granted? We may think of cats simply as pets that live in our homes, which we feed and stroke and are a normal part of our lives. We see their role as 'pets' as being something they are made for. However, the more we look at cats, the more we should be amazed. A cat is a complex thing! The fact that they can live with us as successfully as they do is more of a credit to cats than to people. The more we understand cats, the more we realise how much we don't know and the more we want to know.

Let's first look at cats as cats, not as part of a relationship with us but just as the fascinating creatures they are. This means putting aside our opinions on hunting, on loving and cuddling, on how we live with our cats or what we see cats doing on social media, just for the moment. Keep in mind the word 'adaptable' because later we will come to realise that the cat navigates a dichotomy between the core drive of natural and innate behaviours (mostly developed before cats even came to live alongside people) with an ability to live near people that may actually be at odds with many of those instincts.

How did cats become a normal part of our lives?

Our domestic cats are descendants of African wildcats, sandy-coloured cats with some tabby stripes that come from the Fertile Crescent, a region located today in the Middle East around the eastern end of the Mediterranean, stretching from Turkey to northern Africa and eastwards to modern-day Iraq and Iran. These cats lived a pretty solitary hunting lifestyle, except when reproducing and raising kittens. Human civilisation was changing from moving around with their herds of animals to farming grain crops and keeping livestock. These solitary cats were prey as well as predators and would have been wary of larger animals and people, reacting quickly to remove themselves from dangerous situations or anything they thought could be a threat (which is a lot when you only weigh a few kilograms).

As people grew food around their settlements, storing grain and farming livestock, rodents took advantage of the availability of food – and cats took advantage of the increased numbers of their prey. Those that were less fearful of people and all the activities that went on in and around the human environment took advantage by moving closer. Better food and shelter helped their survival and they could reproduce more successfully, so less fearful cats were able to pass on their genes to their offspring and so on. This 'domestication' of the cat occurred 5,000 to 8,000 years ago. Many would argue that cats are still not domesticated in the same way as dogs or farm animals such as cattle or sheep. We may acknowledge that we 'use' farm animals for our benefit but feel that our relationship with cats is less controlling and more mutual – more on this in Chapter 7.

Of course, at that time, many thousands of years ago, people also realised the worth of having cats around to kill the rodents that ate the human grain and food. They may well have also appreciated the beauty of the cat and fallen in love with the charm of kittens (who wouldn't?). They would have developed distant relationships with those cats that stayed around, but they may also have taken an interest in kittens, using that very early period in a kitten's development when they are less fearful and more likely to form relationships with other species to start to develop closer human-cat relationships. We know that the Egyptians revered cats, probably for all of these reasons, and admired them for their ability to reproduce – indeed they were so valuable to them that they reached God status.

Cats are likely to have lived close to humans for 10,000 years, and even if the domestication has not been as strong or effective (from a human point of view) as with the dog, tameness is thought to be able to start to be genetically expressed within a few generations. However, there is a difference between 'taming' and living alongside us as pets. Cats retain a huge slice of their independent selves and, even today, most domestic cats remain self-sufficient if necessary, and continue to be efficient hunters, even when provided with food.

Retaining wildcat characteristics

Let's look at some of those wildcat characteristics which are still very much part of our domestic felines. We need to understand that cats look like they do and behave as they do because they are hunters and top-of-the-chain predators (in terms of the small prey which they catch). Mother Nature has honed their development to enable them to survive by using their senses, their wonderful athleticism and their curious minds. They have to be able to do this on their own because they are not cooperative hunters and (in the wild, without the contribution humans now make to their diet) they have to be able to catch and kill a number of small prey a day in order to survive. We also need to accept that cats have evolved into amazing hunters and this shapes much of their behaviour

– they are not just furry creatures that sit on our sofas and ask for food! We need to realise that cats are what are called 'obligate carnivores', animals that need specific nutrients which they can only get from eating meat. So while we may have our own beliefs as regards what we choose to eat, taking an animal which looks and behaves like it does because it is such a successful hunter, and trying to make it a vegetarian or vegan, is not only ignoring all of its behavioural instincts and needs but also risking its health.

When cats have to survive without any help from people, they have to carve out an area large enough to provide themselves with enough prey to enable them to survive. Competition can, of course, come from other animals, but cats can at least deter the direct competition of other cats by being very territorial. Their deep-rooted instincts are to defend an area from others, marking it in various ways to show it is their territory and to deter others. More recently in their history, cats have been able to live in groups of related females and their offspring if there are enough resources (food and shelter) for them to survive on. But they will still defend against other strange cats joining and taking their valued resources.

Because of the need to defend their territory, cats developed methods of communication which were more about keeping other cats away than encouraging them, except of course when it came to reproducing, if only briefly. For female cats, the territory had to be able to support the rearing of kittens. For males, a wider area that gave access to female cats while keeping other male cats at bay and provided hunting opportunities for survival would be patrolled and defended. The life of such a male cat was likely to be intense, quite dramatic and pretty short. Female cats would have had less violent lives but had the added stress of successfully providing for kittens as well as for themselves. Of course, many cats around the world still live like this or in related groups away from people.

So what physical attributes does this little hunter need to survive? Having the ability to hunt successfully requires finding prey, outwitting prey, catching prey and killing it with lightning-quick reflexes, all without getting injured. At the same time, they need to avoid danger because the cat is also prey for larger animals. Our little cats cannot quite walk about with the confidence of lions or tigers who have few animals that will challenge or prey on them.

How do cats experience the world?

When trying to understand how and why members of a completely different species act as they do, an insight into how they see their world can be very enlightening. We will have to take ourselves as a reference point, note the differences and use our imaginations. Many animals have truly extraordinary senses and abilities compared to our own, but we don't live very closely with most of them. But our pet cats are part of our lives and part of our families, so we owe it to them to try and understand how they experience the world around them. And, of course, cats did not evolve to live in a human home, so perhaps it can also give us insight into some of the things they do instinctively.

Seeing and hearing the world

We tend to assume that everything revolves around how we tall, rather stiffly upright and slow-moving humans move through our world, with good colour daytime eyesight, hearing adequate to receive the sounds of other humans and a relatively poor sense of smell (compared to many animals). Although they live in the same physical environment as we do, cats experience it in a very different way.

One of the cat's most amazing attributes are its large and beautiful eyes, and the beauty of those eyes is one of the reasons we are so attracted to cats. The eyes are relatively large in a round face, which looks quite baby-like and appeals to us too. However, although the structure of the cat's eye is similar to that of many mammals, it does have its own peculiarities and specialities. It appears cats might not be able to see colours as we do; rather they experience more muted colours of blue and green and perhaps reds as dull greys. Think about when cats hunt in poor light. They are crepuscular (dawn and dusk) hunters, meaning they are active at low light when much of their prey are also active, hoping to be hidden under the cloak of darkness. The large eyes of cats have huge pupils which can open very wide to allow light in and, conversely, narrow to small slits in very bright conditions. At this point the beautiful colours of the iris, which can range from blue to green to amber, become very obvious. A reflective layer within the eye that makes the most of available light, and extra light-sensitive cells on the retina, mean that cats can see well in what will be to us very low light conditions; apparently, they are able to see in light six times dimmer than humans can.

Although cats may not be able to see the same fine detail as we can in daylight, and cannot focus so well on nearby objects (they see best at about 2–6 metres), when it comes to following movement, the cat does not miss a twitch. Special nerve cells in the cat's brain respond to the smallest movement, allowing it to notice the presence of small creatures in the undergrowth – an obvious advantage to a hunter. The dog has always been thought of as the keen-eared pet, but when it comes to sound-sensitivity, the cat can hear sounds of even higher frequency than the dog, and at much higher frequency than people, which means that they don't miss the high sounds made by small rodents, its primary prey.

But just hearing sounds is not enough: the cat has to be able to pinpoint the source if it wants to catch the creature that is making those noises. Cats' ears are highly mobile and can turn independently of each other, allowing the cat to scan and collect sound from all around without even moving its head; sound is funnelled into the inner ear, informing the cat as to where a sound is coming from with great accuracy. The cat can then move swiftly and directly and not rely solely on sight to find the prey's position. Their lightning-quick feline reflexes, combined with an ability to judge distance very accurately, allow cats to pinpoint their prey very accurately and reach it with devastating speed. If the prey freezes – a ploy many prey animals use as a defence strategy – the cat may lose sight of it. However, the cat has immense patience and may simply sit and wait for more movement in order to pick up the animal's whereabouts when it thinks the coast is clear.

Feeling the world

Similarly to other mammals which respond to touch and pressure, temperature and pain, the cat's body is covered with sensors. Most humans are not in touch or in tune with their environment, but the cat is surrounded by what we can imagine as a sensory 'forcefield' that they can switch on by using their coat, whiskers, paws and nose – hairs sensitive to tiny movement and paws sensitive to vibration.

If we draw a picture of a human with parts of the body larger or smaller in relation to how touch-sensitive those parts are, the distorted image (called a homunculus) has large hands, lips, tongue, genitalia and relatively small back, legs, feet and arms. So, a fingertip or tip of the tongue can locate a tiny splinter in our skin while other areas are much less sensitive. When you draw the same touch-sensitive representation of a cat (what would be called a felunculus), the creature also has a large head (especially the tongue and nose) and huge paws. The pads of the paws are very sensitive to touch and to vibration (perhaps the reason many cats do not like their pads being stroked or touched). Funnily enough, at the same time as being so sensitive to touch, the paws are relatively insensitive to hot or cold things and cats seem to be able to walk over very hot ground without worry. The nose and upper lip are the only parts of the cat's body that are very sensitive to temperature and they are used to estimate the temperature of food and the environment. Tiny kittens will use their sense of smell to follow the scent of their mother, but they also follow the increasing temperature gradient with their noses to bring them closer to her. The rest of the body seems to be much less sensitive to heat (at least on our terms) as cats seem to be able to sit on very warm surfaces such as radiators or next to hot fires. Perhaps their origin in hot countries means they can withstand heat very well.

Have you noticed that if your cat is resting and you put your hand nearby, but not touching, the cat knows it is there? They also know (and don't like) if the hand you stroke them with is wet. Far from being an inert layer over the skin, the cat's fur and hair are almost super-sensitive and give it far more information about its surroundings than we, with our layers of inert clothes and fairly insensitive hair on our bodies, can imagine. The cat's coat is highly sensitive and areas of skin rich in touch-sensitive nerves.

Adapted hairs develop into whiskers, not just in rows on the upper lip but also above the eye and on the chin. Similar coarse hairs called vibrissae are found on the elbows and are detectors that alert the cat's nervous system as to what it is touching or movements of air around its body. This helps to guide the cat as it moves in low light with its eyes focused on movement ahead, which all feeds back to silent movement, alerting it to danger and building up a three-dimensional picture of the environment. Whiskers and hairs along the cat's lips also allow it to feel the position of its prey in the mouth and aid orientation to kill and eat. So while we may see the coat as something beautiful, soft and lovely to stroke, it plays a huge part in the workings of the cat's amazing sensory system.

Smelling and tasting the world

If, by now, you are beginning to step into the world of the cat, informed of its environment through its perceptiveness of fairly drab colours but brilliant vision at twilight, its 'seeing' whiskers, foot pads sensitive to touch and vibrations, and a coat which is highly sensitive, you may be ready to be launched into an even stranger world: that of the cat's sense of smell. Imagine cats moving through a sea of scents – like swimming through a sea of colours and textures giving information about the surroundings – past and present. While our own senses of sight and hearing are fairly good, when it comes to smell, we are poorly equipped in comparison and so may underestimate the effect this sense has on our cats.

Cats have the same type of smell detection cells as we do in the lining of the nose, with which they identify airborne substances. However, they have a folded membrane which, in that tiny nose, occupies a greater surface area than in ours and has many more cells, giving the cat an amazing sensitivity to smells. Cats don't follow their noses in the way dogs do when hunting (although they will be able to identify prey by their scent); they stalk using hearing and sight. Their sense of smell comes into play more as a means of communication, to read the messages and marks left by fellow felines and also by members of their 'family', including people and other pets – more on this later.

Taste receptors on the tongue enable the cat to recognise substances that it licks, laps or chews. The cat's tongue functions as a comb as well as an organ of taste. Down the centre are backward-facing hooks which help to hold prey as well as being used to lick food or remove tangles from the coat. Cells on the tongue are sensitive to temperature and taste. Cats do not have a sense of 'sweet', although some cats do seem to like food we would class as sweet, but even then they may not be experiencing the same sensation as we do. Since cats are obligate carnivores (they need the constituents found in meat in order to survive) this is not surprising as their taste buds are aligned to be stimulated by chemicals that are constituents of proteins. Fats in different meats probably also smell and taste different.

Cats also have an additional sense which combines both smell and taste. When smells (chemicals in the air) are trapped in the mouth, the cat presses its tongue against the roof of its mouth to force the air into a thick tube of cartilage about 1 centimetre long. Situated above the roof of the mouth and opening up just behind the cat's front teeth, this is called the vomeronasal organ or Jacobson's organ, and it seems to enable the cat to concentrate smells and taste them at the same time – an additional sense which we humans have lost. However, it is not necessarily food smells for which the cat uses the vomeronasal organ to test – its primary function is to pick up what are called pheromones, which are chemical signals produced by, for example, female cats in season.

Pheromones affect and automatically change the behaviour of another animal of the same species. They are similar to hormones, but hormones work internally and only affect the animal secreting them – pheromones are secreted outside the body and influence the behaviour of another cat. Many of the messages being sent are about the reproductive status of the animal, but we also know there are pheromones that communicate other messages such as fear or stress. When smell/tasting these pheromones, the cat exhibits a strange behaviour: it stops, stretches its neck, opens its mouth slightly and curls up its top lip to draw in air to the right place. It is quite subtle in cats, so you have to watch carefully – but you may have seen horses doing this, where you can see the mouth and lip movements much more easily as the horse raises its top lip in a rather comical way. This grimace-like face is called the flehmen response and can be seen in male and female cats, neutered or entire. It is also seen in response to catnip or catmint.

Moving through the world

Using its heightened senses of hearing, smell and touch, the cat may be influenced by factors of which we are only vaguely aware. Built as a perfect hunter, its senses are attuned to help it move swiftly and silently through its world. A cat's amazing claws are retractable and are brought into play for hunting, climbing or defence, but left sheathed for silent walking. So, with all this data coming from its senses, it has to be able to act swiftly and decisively – a cat's body needs to be fit, agile and faster than its prey. Its compact size and shape, balanced by a tail which, like many of the cat's physical attributes, has both physical and communication uses, allows the cat to move not just along the ground, but to climb and jump easily to higher levels. Supple and rarely clumsy, the cat has a physique that enables it to perform feats way beyond those of a human athlete. A cat can move within its world with seemingly little effort, leaping to many times its height, climbing, jumping and balancing with agility and confidence. That isn't to say accidents and falls don't happen, because they do, but the additional ability to turn in mid-air to fall on all fours means that cats often survive better than other animals would from a similar fall.

Communicating with other cats

So far we have looked at the attributes of cats in the context of how they hunt and survive. But many of the cat's physical and sensory characteristics have a dual purpose. The agile body and super senses are not just used for hunting and avoiding danger – they are used in communication with other cats. This is the next step in better understanding our cats: what body language and behaviour do they use in communicating between themselves? How do they pass on information to other cats?

Consider the lifestyle of a cat that is not living with people – it is territorial, defending a territory in which it has the chance to hunt for enough prey to survive. If there is plenty of food, then a number of cats may congregate in this area. You may have seen street cats or farm cats living in groups; in a breeding group of cats these are likely to be related female cats and kittens. However, they will still defend the area to stop other strange cats coming in. The methods they use to do this are to leave messages (pages 26–7) to keep other cats at a distance so that they do not have to interact with them or repel them physically (which may cause injury).

The exception to this is when female cats come into season and are seeking mates and, in this case, the messages they send are clear and welcoming. Within a group, cats communicate to verify the group members that are not a threat to them by exchanging individual scents and building a group scent profile. They do not cooperate on hunting or defence, but may share the care of kittens. They also use body language, some of which is subtle (in that we as humans may not notice it every easily although it may be very obvious to cats!) or much more obvious and dramatic.

Exchanging scents

When cats that are familiar and friendly to each other meet, they rub head, flank and tails against one another, exchanging odours and greetings, just as we would shake hands or kiss a friend and make light conversation. Their straight-up tail stance is a clear and strong signal that the cat they are meeting is a friend and is welcome – they may then approach each other nose to nose. It also allows them to investigate each other's anal region, where glands are situated under and above the tail – just to check that the scent matches. Grooming each other also helps in the exchange of scents and to develop a familiar group scent that is reassuring to them.

However, many messages are aimed at leaving information which can be received from a distance and with the aim of keeping other cats away. Cats leave scents around their home range or territory for other cats to find. By rubbing lightly against twigs or other objects, they leave a smear of oil-based secretion containing their scent. They may also deliberately rub the chin/lip area against the end of a stick to anoint the tip with secretions from the glands around the mouth. Other cats take a great interest in these scented objects and may move from one to another, often over-marking them with their own scent as they pass. By investigating each one they can tell when other cats passed and leave their own message. Small gaps in fences and hedges through which cats squeeze when on patrol also become smeared with body oil and scent and give similar clues to the cat's passage and occupancy of a home range. Cats may avoid a territory totally or may 'timeshare' their ranges, moving around at different times so they can avoid meeting.

Cats have sweat glands over the body, but only those on the pads of their feet are similar to our own. These secretions keep the pads moist and prevent cracking and flaking (as seen on the pads of dogs' paws) so that the pads remain sensitive and flexible. Sensitivity is especially important during night-hunting when cats need to be able to feel what they are walking on while keeping eyes firmly fixed on their prey.

Three things happen when the cat scratches an object – it leaves a physical and visible mark, it leaves a scent message, and it removes the old nail husks to reveal new points vital to keeping its hunting weapons sharp. The scratching also seems to have a use beyond these three things – almost showing off or attracting attention – something to remember when we look at how cats behave with us and what we might gather from these behaviours.

Animals also use urine as a message carrier that can be left and positioned for maximum impact. Urine can carry a number of markers which tell others about the cat, such as whether it is in season or is an 'entire' male cat marking the territory. The urine is often delivered to be sniffed at cat nose height, with the cat turning to face away from the wall/trec/ bush on which the scent will be deposited with its tail held high; a quivering motion of the tail may be accompanied by a paddling or treading motion with the back feet. The fine spray of urine is squirted backwards onto the vertical surface.

Depositing the urine up off the ground also allows the scent to be carried on the wind. Apparently, tom cat urine (which smells very strong even to our insensitive noses) can be detected by another cat from 12 metres away and can last for up to two weeks, depending on the weather. Certainly, for days afterwards other cats can tell a great deal about the sprayer. The chemical markers decompose at a certain rate (depending on the weather), which means that a visiting cat knows how long ago it was deposited and can gather information about the sprayer. The spot will be topped up as the scent degrades, or other cats, too, may spray on top of it.

Body language

Cats are not socially obligate like dogs or people – they don't need to have others around them to make them feel secure. Cats may prefer to be solitary or can be a bit more flexible if the circumstances and the other cat(s) fit, indeed they may be much more relaxed without them around. Thus, cats have not developed with a lot of overt social communication skills like people or dogs because they do not collaborate or compromise to keep individuals together. Those behaviours that are designed to be seen from a distance are larger and more exaggerated, while those designed to communicate at a closer distance can be quite subtle and difficult to notice. But, of course, they may be obvious to other cats, and more on this later.

If you want to turn up the dial of body language so it

becomes more obvious to us humans, you should watch how kittens behave or when cats play. Everything is exciting or scary for kittens and their emotions are expressed very clearly with exaggerated body language. Similarly, body language between rival unneutered males can be noisy and easily recognisable – the aim being to intimidate and chase off rivals without resorting to actual fighting. However, they often start with a staring match, which may not be obvious to the human observer but can be powerful in cat language.

In a moment we will look at some of the detailed body language of the head, tail, whiskers, ears, eyes, etc. While these all have a function for physical survival, most have a dual function in showing how the cat is feeling. However, taken in isolation they may not tell the whole story, and the emotions that the cat is expressing may be difficult to tell apart and are easy to misread – we'll look in detail at what our pet cats are telling us later. To get the whole story, the entire body must be taken into account. Isolating one feature may also be misleading because signals often change rapidly as the cat's mood and mind alters and a situation changes. The cat may be trying to look small and hide away from something threatening, in which case the limbs and tail are drawn in and the body and head lowered, ears usually lowered, and pupils dilated and avoiding eye contact. A cat trying to look large and threatening will increase the impression of its body size with hair standing up on end and head turned to stare at a rival. A cat on the receiving end of a threat may turn sideways with raised hair to try and look large enough to frighten off the other cat while it very slowly creeps away without triggering an attack, or it may try to shrink to be small and hidden. Of course, these are rather dramatic, and lots more happens in less extreme situations, with body postures more subtle.

Cats are great watchers – to find prey, to decide if something is friendly or dangerous, to predict activity, to learn, or just because they are curious. Kittens learn by watching their mothers and their siblings. But the eyes, relatively large compared to their body size, emphasising the importance of sight, are also great at conveying the cat's emotions. Emotions like excitement or fear cause the release of chemicals like adrenaline that affect the size of the pupil, which can go from a slit to filling almost the entire eye. So, the eye may be wide open with a dilated pupil because it is dark or because the cat is frightened. And it is not just the components of the eye itself that tell a tale. Cats use a direct stare as a way to wrestle and threaten each other without resorting to physical interaction which can result in injury, so they are sensitive as to how they look at each other.

Cats' ears can swivel through 180 degrees and can move independently of one another to follow and pinpoint sounds, but they have an emotional as well as a functional use, able to convey feelings such as interest, fear, anger, confusion and frustration. Some of the big cat family have additional markings on the back of the ear, presumably to exaggerate the movement or message they are conveying; others have tufts on the top of the ears. Pricked forward-facing ears show interest and comfort/confidence. Other obvious positions may show a 'u' shape, with the ears rotated, which seems to show fear or discontent or confusion – the cat is aware that something is not quite right and is considering what to do. Ears can also be flattened against the head to avoid contact or make the cat less noticeable. Of course, there are many ear positions between these extremes and one ear may do one thing while the other does something completely different. They can change very quickly and may be swivelling to pick up sounds as well.

A happy, relaxed cat will usually sit with its ears facing forward but tilted slightly back. When its attention is caught by a noise or movement, its ears will be pulled more upright and become more 'pricked' as the muscles in the forehead pull them in. It is rather like us wrinkling our brow when we concentrate. If the ears begin to twitch or swivel, the cat is probably feeling anxious or unsure of a noise or situation. If anxiety increases, the cat moves its ears slightly back and down into a more flattened position.

As we have seen, whiskers have a functional use when the cat is moving in low light or in helping to work out the position of prey, but they too can give clues as to how a cat is feeling. Whiskers are much more mobile than we imagine and can be fanned out forward in front of the muzzle or be pulled in and held back against the cheeks, both moving them out of the way or forward to touch something, as well as conveying excitement or interest.

The cat does not use its mouth in aggressive confrontations in the same way as dogs – a cat may growl without mouth movement and a cat's open-mouthed hiss can be used to respond to threat. Licking of the lips may be a sign of anxiety but should not be confused with cats that sometimes sit with their tongues sticking out, which just seems to be a relaxed moment (one we often find quite amusing!).

Open-mouthed breathing or panting is not a sign of overheating in cats but should be acted upon as this may be a sign that the cat is having difficulty breathing. Yawning, which often accompanies a languid stretch after waking, is thought to be a sign of reassurance and contentment in cats. Of course, cats do use biting when defending themselves and a cat bite can be a serious injury – it is often deep because the long narrow canines cause a deep injury with a small entry wound that may then close over. Bacteria is then captured in a pocket which may develop into an abscess. A cat bite in a person should always be taken very seriously and treated with antibiotics.

Like all the other parts of the cat we have looked at, the tail has a functional and a communication role to play. When hunting, the cat folds its tail in a streamlined manner behind until it is required as balance for the final rush at prey. It may also announce the cat's interest and concentration with a twitching movement as it watches its prey. The tail is mobile, being able to sweep from side to side and up and down, slowly or very rapidly. It wraps sleekly around the cat as it sleeps but can turn into an erect bristling brush if the cat is very frightened. When meeting a cat or person with which it is friendly, a cat will flick its tail into a vertical position, pulled slightly forward over its back and kinked down a little at the tip. A kitten greeting its mother will run up to her with its tail in the air and proceed to let it drop over the mother's rump and rub over the top of her tail in an attempt to solicit some of the food she has brought to the den. So, tail-up is always a great signal of welcome.

As kittens play, they often use exaggerated movements that you rarely see in adult cats unless they are in extreme-emotion states or during play when they seem to revert to the openness of kittenhood. The 'inverted u' shaped tail can be seen when kittens play or when cats have a mad half hour and chase each other. It is probably that feeling of a mixture of fear and excitement which children love when they are playing a hiding game or being read a scary story! In general, the tail will sweep slowly and seemingly haphazardly from side to side when the cat seems to be idling, considering what to do next. Watch two cats at play – it can often spill over into more of a rough and tumble and may end in some hissing (my mother used to say 'it will end in tears' when we kids got too excitable when playing, and this does seem to fit the cat-play scenario sometimes). You will see the tail posture and speed of sweep varying – it is a part of the whole-body posture and movement. Violent thrashing indicates high emotions. These behaviours and movements are general –

individual cats will use their tails differently in everyday life, and interpreting how your cat uses its tail will be looked at in Chapter 9. For now, understand the tail as a tool for communication – the faster it is moving or the more it is fluffed up, the more agitated the cat is.

That's a whistlestop tour of some cat body language and behaviour – it gives a background against which to get a feel for the species. We are probably unaware of many of the more subtle signals cats give each other, but we are gradually getting an idea. The title of this book is *How to Talk to Your Cat* and, borrowing International Cat Care's first 'Cat-Friendly Principle', learning to respect the species and then getting to know the individual cat will help you to understand your cat and how it is with you, with all of this behaviour knowledge in the background.

Sound communication

The word 'talk' for us is, of course, the important part of communication, but when we talk to someone, we are also reading their body language and how they are responding, or with some knowledge of the person they are. In this book, 'talk' is a broad term covering how you approach and communicate with your cat.

So far, we have not looked at the sounds cats make – the very narrow definition of 'talk'. Considering how much store we set in vocal communication and that cats are such popular pets, we know or understand relatively little about the sounds they make. Cats are said to have more than twenty sounds they produce, and you could group these into the types of situations they are used in.

Cats behave differently with people than they do with other cats, so first of all let's look at cat-to-cat 'talk'. Cats use vocal communication with other cats in three main situations – mother and kitten, reproduction and combative. And if we add humans to the equation, there are cat-human noises too. The noises cats make vocally can be loud or quiet and can be combined, repeated or produced at the same time. Cats form noises in a slightly different way to people: the tongue plays a less important role and sounds are made further back in the throat. Air is pushed at different speeds over the vocal cords stretched across the voice box, changing the tension in the throat and mouth muscles, which changes the quality of the noises. Cats use different sounds in different situations, and this can be very individual – especially where mothers are interacting with kittens. The same communication sound, such as the mew or the purr, may be used in different places, as we will see.

So let's look at the ways cats communicate with each other and then go on to think about the way they communicate with us, because that is something additional to their natural repertoire. There are two areas of vocal communication that we don't see (or hear) very often in our cats. These are the sounds used during reproduction and when they are in conflict (at least we hope we don't hear these too often in our homes). Then we can think about noises which are friendly around cats (and people) and those which are used when they want to do something or encourage interaction.

Reproduction and conflict sounds

During reproduction female cats will 'call' – making loud and drawn-out meow/moan-type sounds which encourage male cats to come and find them. There may be strong competition between males to find a female cat in season and male cats may caterwaul or howl with long, loud drawn-out sounds to signal their presence to both female cats and competitor males. If another cat gets too close, the howl acts as a warning to avoid fighting. The howl is often combined with repeated growls and howl-growls! The sounds can also be used by female cats as a warning. Female cats also emit a low, quiet growl while mating, and may emit a cry at the end of mating and often turn aggressively on the male. These sounds can sound quite combative!

Other sounds used in exchanges that are not friendly include hissing, spitting, growling, snarling and howling, used by the cat to warn, threaten or shock. Growls (rumbling noises) are used to warn off another cat it doesn't like from coming too close and are made with a slightly open mouth. The growl can be used along with other sounds such as hissing. A hiss is produced by pushing out air quickly through the open mouth with teeth showing and is often a reflex to surprise that frightens the cat. A spit is a more intense type of hiss, a short explosive noise made by breathing out violently to warn or scare off, an automatic response to something frightening.

Cat-to-kitten communication

Sounds cats make to contact each other include a lot of those used between a mother and her kittens. Some of these sounds they use with people too; it's lovely to know they automatically use these interactive and encouraging sounds with us. Although kittens are born with their eyes shut for about ten days and their hearing does not function fully until they are about four weeks old, they can make noises from very early on for their mother to hear.

Researchers have found that kittens use very specific sounds to communicate to their mother if they feel isolated – presumably when they are away from the nest or just feeling alone, perhaps not snuggled up with other kittens. The sound is also individual to the kitten, so the mother can identify her own kittens and, in the same way, kittens can recognise their mother's 'chirp', which reassures them as she comes to the nest. She uses the same sound to encourage the kittens to explore out of the nest with her as they get older. A chirp is a short high-pitched call named because it sounds a bit like a bird or a rodent chirping and, if several are used together, they are called chirrups. She also makes a 'gargling' type sound when she brings prey for the kittens, and this may vary with the prey so they are prepared – perhaps, for example, if it is a rat and more dangerous than a mouse. It's a fascinating part of cat behaviour and there is probably lots more to learn about this aspect of cat communication.

Purring is also a means of communication between mother and kittens, which calms them. Kittens can purr back to their mother, telling her all is well. The sound does not carry too far, which means it does not alert predators. It can encourage the mother to feed the kittens and the kittens to feed from her. The kittens also use purring to encourage other kittens to play or to do something they want (remember this when we look at our relationship with our cats). There's more to the purr than this, but we will consider this when we consider human/cat interactions because cats can use different purrs to encourage us too!

Adult cat-to-cat vocal communication

Interestingly, cats seldom make miaow-type noises with each other – perhaps they don't need to. Researchers say that our domestic cats have a more developed and complex range of vocal communication than their wildcat cousins. In truth, we know relatively little about the sounds cats make to each other and what they mean and, indeed, scientists guess that our little domestic cats do not make a great deal of noise in case they attract attention from predators bigger than them that would view them as prey.

Other softer sounds such as the murmur are produced with a closed mouth during friendly approach and play, and cats do make little noises to each other in what seem to be positive interactions. Scientific papers also list trills, tweedles and tweets as sounds cats make, but I must admit I have not actually managed to identify these individually or to put the right name to each one!

Cat-people sound interactions

Scientists have found that feral cats (which live and reproduce in the human environment but away from people and pretty independently) produce higher growl and meow sounds than pet cats that live with people, which may indicate that they live in a more stressful and fearful situation. This isn't exactly surprising for such cats living on their wits. Street cats that live in groups don't often meow to each other. In studies of feral cats which are fed by human caregivers they found the cats didn't meow to them, but apparently vocal communication did increase as those cats got used to people.

Cats have evolved to use the sounds they naturally make with other cats in interactions with people and have adapted how and when they use them, changing the intensity and tone of the noises. The mew, meow or miaow is used rarely between cats but is the sound cats use most commonly with people to attract attention, often when they think they are going to be fed or when asking to be fed by their owners. It has many uses and is made with the mouth starting open and closing slowly.

Cats can vary the sound of the miaow and we can interpret

that and even notice if it is a positive or negative miaow – and amazingly they can make it just for us. They are clever creatures and read us well – when we react to their sounds, they learn what works best and adapt the sound. It's a great survival tool.

What is the sound we most love about the cat? The purr. Kittens purr at an early age when they breathe in and when they breathe out. Purring is done with a closed mouth and can occur simultaneously with other vocalisations. The sound and vibration are produced in the larynx by changing the flow of air. The diaphragm and other muscles appear to be unnecessary for purring other than for breathing, which actually increases in rate if the cat is purring.

Cats can keep up their purring rhythm for hours, the sound varying in loudness from the rough purr with a distinctive 'beat' to the smooth, drowsy or almost bored purr with an indistinguishable beat that suggests it will probably stop soon. A higher-pitched purr is often used when the cat is eagerly seeking attention. Two different types of purr have indeed been distinguished in cats: the purr produced by cats that are relaxed and contented, and the 'solicitation' purr produced by cats actively seeking food or attention – the sound of this purr is perceived as more 'urgent' by owners.

A study reported in a journal which publishes research about sound showed that cats could purr at frequencies optimal for pain relief and even tissue repair. Purr frequencies correspond to vibrational/electrical frequencies used in treatment for bone growth/fractures, pain, fluid accumulation, muscle growth/ strain, joint flexibility and wounds. An internal healing mechanism would be advantageous, decreasing recovery time and keeping muscles and bones strong when sedentary.

Adult cats purr to one another in the same reassuring way or to encourage a reaction such as grooming from another cat, and they may use purring for self-reassurance when in pain or to placate a potential nearby aggressor. Is the cat using the feeling of wellbeing associated with feeding or positive interactions to reassure itself when in pain or fear? We probably don't know. They use their purr frequently with people and we love it! We'll explore this more in Chapter 8.

Other noises

Cats also make some strange sounds which are not directed at people or other cats as far as we can ascertain, and have no role in communication but perhaps signal frustration or pent-up energy. The strange tooth chatter which is often produced when a cat sees something it wants but can't get to, such as a bird outside a window, is made by the mouth being slightly open, lips pulled back and the jaws opening and closing rapidly, clashing together and creating a smacking sound. The noise is a combination of lip smacking and teeth chattering as the cat gets more and more excited. The cat may also emit small bleating/nickering type sounds as its teeth chatter.

Cats may also make rather extraordinary sounds when bringing prey home – it is quite loud and guttural and repeats several times. If you can't see it, you may wonder if a strange cat has come into the house. It seems to be a way of signalling that the cat has come with food, perhaps used when kittens are larger and the mother cat is teaching them to hunt and how to deal with prey and calling them to her. We know when we hear that noise in the house that we are likely to find a mouse on the floor!

So, these are the basics of how cats see the world and how they communicate with each other using body posture, behaviour and sound. Have them in your mind as moving on they will help us to understand our cats' behaviour with us.

2.

What affects how cats behave around people?

SO FAR, WE have looked at cats as if watching a David Attenborough wildlife documentary, where the camera person stays out of the way and the audience tries to deduce what is happening without the interference of people. But we are primarily interested in cats because we have them as pets, as part of the family, and have an emotional bond with them. There are millions of cats around the world, and they exist because of people, the shelter and food we provide (either accidentally as rubbish which can be scavenged or by actual feeding) and our desire to have cats around. We have a responsibility to care for them however they live.

What happens when we add people to the cat equation? How do we affect cats and how can we understand them better to ensure we care for them as best we can?

How do we live with cats?

Let's first look at the range of lifestyles cats have and how these are interwoven with people.

There are few places in the world where cats don't come into contact with people and survive alone. Not every cat is a pet – think of those you see when you are on holiday somewhere like the Mediterranean, where they survive on the streets. International Cat Care has a great diagram (see below), which shows the range of cat lifestyles. This helps us understand why not all cats are the same and why some are more able to live with us than others. In this understanding we can learn how to help each of these cats in a way that is respectful of their lifestyles and experiences.

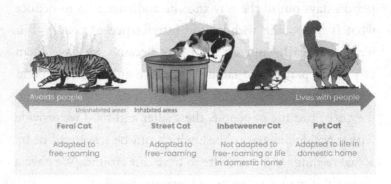

Picture courtesy of International Cat Care showing the different lifestyles of cats according to whether they are adapted to being independent free-roaming cats or are adapted to live with us as pets.

Let's have a look at the different categories. A feral animal is one that lives in the wild but is descended from domesticated individuals. Many cats are called 'feral' just because they live on the street and it is a word used differently all over the world, usually just to suggest that a cat is not a pet. However, those we would say are truly feral live in uninhabited areas; their lifestyle is what we call free-roaming, in that they come and go as they wish, living independently and without input from people. They have probably been born to other feral cats and live outdoors and hunt to survive. They are many generations away from cats that have lived with people, are not used to them and would not want to be confined in a human home.

Street cats are those we often see on holiday, in the Mediterranean for example; they live outdoors, survive by hunting and scavenging and are often fed by people, too. They may live in groups around areas where food is found, such as near hotels or ports where it is available to scavenge or where they are fed by people sympathetic to them. They may seem friendly in that they come close to feeders or to people eating outdoors at restaurants, but probably won't want to be handled or confined. Indeed, they are likely to become very distressed if forced to live in a human home.

Pet cats live with us in our homes, sometimes entirely indoors and often with access in and outdoors. They enjoy our interactions and are used to people; they even relax with unknown people. If a pet becomes lost or abandoned, we may call it a stray cat.

When I was working at International Cat Care closely with Vicky Halls, we thought a great deal about the cats which were in homing centres (also known as rescues, adoption centres or sanctuaries). Some are friendly and interact with people when they come in; others hide all day under blankets or in the litter tray. When cats first come to a homing centre, most will be frightened by new sights, sounds and smells and will try to hide away. However, some soon emerge and interact while others seem to be stuck in anxiety and fear. We gave a new name to these cats – 'inbetweeners'. These cats have previously lived as pets in that they have been cared for in people's homes. However, the relationship has not been very satisfactory because these cats were very uncomfortable around people. It is likely that they did not have the right quality or quantity of interaction with humans when they were young kittens (see later in this chapter) and are fearful, anxious and distressed around them as a result. Owners of inbetweeners often describe how their cat hides away from them or is quiet (or scratches, bites or hisses and growls if they try to handle it or pick it up) and disappears when other people appear. The cat may spend lots of time outside and come in for food and water and warmth. It may urinate or defecate in the house because it is stressed. This is a cat uncomfortable in a human environment and may be a disappointing cat for owners who want a cat that likes to be stroked or picked up – what we would call a 'pet cat'.

When most of us refer to cats we mean pet cats. In the

dictionary a pet is defined as 'a domesticated or tamed animal which is kept for companionship or pleasure'. This seems to be a very one-sided definition based on the person's need for companionship or pleasure, one that doesn't take into account the animal's companionship or pleasure. We want cats that are happy to be with us and do not need to be 'tamed' to enable them to share space with us. We want them to view their lives with us as one which is mutually enjoyable and in which we are comfortable with each other.

How are pet cats made?

How is it that we have these different lifestyles? Cats are not born wanting to be pets. Kittens have no concept of what being a 'pet' is or what living in a human environment will mean. Perhaps we would define a pet as a situation in which cats and people live together comfortably, both benefiting from the relationship. A cat living with what we might call a responsible owner will benefit from veterinary care and enjoy food, shelter and warmth. People will want to stroke, pick up and even cuddle and kiss pet cats and expect them to at least be tolerant of this, and at best show that they enjoy it. It is understandable that this is the outcome people want for all kittens. However, for some cats, being near people can be highly stressful. So why do some cats enjoy being with people while for others it is a very fearful experience?

Whether individual cats enjoy living with, or would rather

avoid, people is due to a combination of things. These are: genes inherited from parents, the effect of the environment on the mother when she is pregnant (which is likely to impact on the kittens), and the kitten's experiences, first of all in the first two months of life and then its experiences later on.

It makes sense that, as in people, behavioural tendencies and temperament traits in cats are genetically inherited from parents. Some people seem to be of a more nervous or fearful disposition than others and some more confident and bold. If a kitten is genetically more likely to be fearful, then getting it used to life as a pet cat and living with people is going to be far more challenging.

One of the most obvious ways in which we can see how selecting certain genes produces different types of cats is to look at the variety of physical appearances found in different pedigree breeds. The variations are obvious, including body size and shape, head shape, coat type, length and colour. Breeds are maintained by only allowing cats with certain characteristics to mate with other similar cats or by introducing variations in a controlled manner. Thus, there is a good chance that, by using this selective method in the mating of cats, the disposition for certain behaviours may be passed on to the next generation along with the physical characteristics. We look more at breeds in Chapter 7.

We are still learning about how we as humans develop and how experiences early in life may affect us throughout our lives. Research has given us some insight into kitten development, but we are probably still missing a great deal. However, it seems that there is a window of time when a young animal is particularly influenced by its surroundings, how it learns from experiences and how it processes that information. Behaviourists have labelled this time the 'sensitive period', and for kittens it is 'open' in the first two months of life. When we think of how quickly kittens have to develop from helplessness to independence, it is not surprising that this is the time when the brain has to learn quickly what is OK in life and what is to be avoided. They are still being protected by their mother at this point, so have guidance while they learn. After this period, they will need to react quickly and automatically to survive. If they have positive experiences with people or other animals during this time, then their responses to these will probably be positive for life, because the brain processes have been laid down and the brain is less plastic or adaptable when this window closes. If they have had a bad experience or no experience of people during this time, then forming bonds may not come automatically, and kittens turn into cats that are fearful of people.

This short period can have a huge effect on how cats deal with life in general and continues to influence them throughout life. Relationships that depend on familiarity are formed at this time – so humans and other species such as dogs may be incorporated into the cat's social group and responded to without fear. On the other hand, if the right exposure doesn't take place, some abilities never, or only partially, develop, resulting in irreversible fear when encountering, for example, people or dogs.

During this critical time, kittens also get used to sights, sounds, smells, touch sensations and activities that go on around them. They learn to recognise what they should be frightened of and avoid, fast learning to lay down responses which can be called on when the cat is independent and having to make rapid decisions to protect itself. If this is successful, it enables a kitten to recognise and become less reactive to harmless things. Kittens that experience a wide range of things during this time and learn to deal with them are able to take this ability to deal with novelty into life in general and it increases their ability to cope with a changing environment, such as in a human home. However, if kittens are born and develop during this period in an environment that does not give them the chance to experience and get used to things commonly found in a human home, then what we want them to find reassuring in our homes may instead cause stress and fear.

All this emphasises the huge responsibility that people who purposefully (cat breeders) or accidentally (because they have not neutered their cat in time to prevent it becoming pregnant when it comes into season) produce kittens have. What they do with and around the kittens in those first two months of life will have a substantial influence as to whether the cat is happy to be a pet and relax with people. Cats feeling stressed at being in human homes can try to cope with the environment by hiding or urine marking, actions often seen by people as 'problem behaviour' but which are actually just natural feline behaviour as they try to cope with fear or anxiety.

Young kittens need to get used to being gently stroked, lifted and held. The aim is that all this attention should feel good to the kitten. It is vital that all handling should be gentle, non-threatening and should feel secure. As with children, it is best to choose times when kittens can concentrate on and enjoy interaction rather than when they are tired or very excited. Research has shown that being handled and spoken to gently for a total of 40 to 120 minutes per day, over multiple short sessions during the kittens' sensitive period up to seven or eight weeks old, helps kittens feel more confident about approaching people and more likely to enjoy physical contact with them. Handling in the presence of a calm mother and littermates will also help this to be a positive experience.

People often find that while their cat may be fine with them, they are frightened by people of different genders or ages – these people are different to perhaps a single owner who interacted with the kittens during this sensitive period. Meeting people of different ages, sex, sizes, etc. helps the kittens to accept differences later on because variety in people becomes normal. Research has shown that kittens that are handled by four or five different people (male, female, adults and children) before they are seven weeks of age will be happier to be around people as adult cats and more likely to interact with them. On the other hand, cats that are only handled by one person may be affectionate towards them but see other people as threatening, and are more likely to avoid them.

Being exposed gradually and gently to as many sights, sounds, tastes, textures, smells and activities as might be encountered in a normal domestic home lifestyle – e.g. vacuum cleaners, washing machines, televisions, different floor coverings and car journeys – will go a long way towards helping kittens to become cats confident in a human home with human activities.

Kittens of course keep learning beyond eight weeks of age and, if they are given a good start, then they will not be fearful and will be able to interact and find out about new things with confidence. If they have missed those early experiences, they may well try and run away or be too frightened to learn. So positive builds on positive and it seems that if kittens miss those early experiences with people and their environment, they may unfortunately never be able to make up for the lack and remain fearful in the human home. Cat experts think that a kitten's attitude to people is likely to last at least three years, but probably for their entire lives.

An additional factor that can affect how well a kitten is able to face the world and its challenges is how stressed or relaxed its mother was when she was pregnant. While it has not been studied specifically in cats, many offspring born to other mammals that are stressed during pregnancy can suffer from what are called 'epigenetic effects'. This means that stressful environmental conditions seem to affect offspring, causing them to react more to stress, which can then make it even more difficult for these kittens to become cats that are able to live with people without fear. Of course, the temperament of the mother and her attitude to people will also rub off on the kittens if they see her interacting calmly and positively with people. Unfortunately, we do not understand the vital learnings which go on between mother and kittens, but when we try to hand-rear kittens which have lost their mother we find it is not as easy as simply providing food. We don't know how mothers communicate how to behave to kittens. It is not uncommon to find that some hand-reared kittens are difficult to live with, demanding and seemingly aggressive when they don't get what they want. This may be linked to learning how to deal with frustration, but we are guessing somewhat.

Do the characteristics of kittens persist into adulthood? If we choose a kitten for confidence and friendliness, will this change as it grows or, perhaps more importantly, if we take on a nervous or fearful kitten, will it grow more confident? Some research has shown that kittens that are inquisitive and active at four months can still be categorised as such at a year old – perhaps linked to boldness or confidence.

If a kitten has had good, early experiences with people it is much more robust about negative experiences and it may take several negative experiences to become wary of people, and as a grown cat is able to become friendly and trusting of a new person. A cat that has not had the right experiences as a kitten and in early life may need a large number of positive experiences before it can accept a new person. However, only a few negative experiences will help it to realise that it was indeed right to be very fearful of people in the first place. And if the cat is fearful and removes itself from people and survives, then it feels it has made the right decision and this reinforces the decision to keep away from people in the future!

So, the kitten's genetics, its mother's environment when pregnant, and the kitten's early experiences all combine in the formation of a kitten's personality and affect how it faces the world and copes with challenges and change. How a cat behaves can also depend on the situation in which it finds itself. A cat that is very easy-going in one situation may react in what seems to be an aggressive way in a different situation – being aware of this may help us to understand behaviours that seem to be inconsistent in our pet cats. And our approach to and behaviour around our cats can have a big effect on their behaviour – more on this later.

We have already looked at how kittens develop in the early weeks and how this affects their approach to the rest of their lives. So far we have seen what makes a cat the personality that it is – from its genetics to its early experiences as a basis for its character. But, of course, that is just the foundation and will affect how a cat tackles the challenges which affect it and the relationships it develops. You may have a basically nervous or fearful cat, or you may have a bold or confident cat which then approaches life rather differently. So, what else could affect how your cat interacts with you?

Age

We have seen how a cat's basic personality is probably set quite early on in its life; however, over the years the cat continues to learn and develop. It can be very difficult to age a cat by looking at it. When is a cat a teenager or when does it enter middle age? Do you multiply a cat's age by seven to compare it to human years in the easy and historic way we think about dogs? That is quite a blunt instrument for thinking about equivalent age – International Cat Care has produced a diagram (see the next page) which shows equivalent ages which have been calculated to take into account how quickly cats develop and when they mature physically and behaviourally. Cats age pretty gracefully; they don't go grey like their canine cousins, although the fur may change colour a little bit, black going towards brown. Also, they retain a lot of their graceful mobility until they are probably in their senior years and can still sometimes play like kittens into old age. Even if they are afflicted by arthritis, which a surprising number of cats are, they cover pain exceptionally well and so do not limp or show pain to a great degree. But, as we will see, there are things to look out for that can give us clues.

For their size, cats live quite a long time. Generally, bigger animals live longer than small animals – think about an elephant and a mouse. Tortoises, man and a few other animals are exceptions to the rules. A small mammal such as a mouse may live for two years or so, a pet rabbit probably averages

HOW OLD IS YOUR CAT?

	Life stage	Age of cat	Human equivalent age
	Kitten Birth–6 months	0–1 months	0–1 years
		2 months	2 years
		3 months	4 years
		4 months	6 years
		5 months	8 years
		6 months	10 years
	Junior 7 months–2 years	7 months	12 years
		12 months	15 years
		18 months	21 years
		2 years	24 years
	Adult 3–6 years	3 years	28 years
		4 years	32 years
		5 years	36 years
		6 years	40 years
	Mature 7–10 years	7 years	44 years
		8 years	48 years
		9 years	52 years
		10 years	56 years
	Senior 11–14 years	11 years	60 years
		12 years	64 years
		13 years	68 years
		14 years	72 years
	Super Senior 15 years +	15 years	76 years
		16 years	80 years
		17 years	84 years
		18 years	88 years
		19 years	92 years
		20 years	96 years
		21 years	100 years
		22 years	104 years
		23 years	108 years
		24 years	112 years
		25 years	116 years

Picture courtesy of International Cat Care

eight years and a dog between seven and twenty depending on its breed or size, its activity, or both. Cats are a little bit bigger than the average rabbit (and smaller than some of the giant rabbit breeds) but will live on average about twelve to fourteen years. It's not unusual for cats to reach their late teens or even their early twenties. We can think of our cats as having six life stages, from kitten to super senior.

The first stage of a cat's life is kitten, but they don't stay kittens for very long – by six months old they have grown rapidly and learned quickly. It brings kittens up to a period when, if they are not neutered, and depending on the time of year, temperature and day length, they are able to start breeding themselves. The rapidity of this development often catches owners out because they do not realise that their kitten can become a parent. Most welfare organisations now suggest neutering our pet cats at around four months to remove the possibility that they become mums at this early age.

During this early six-month period a kitten has to go from being born helpless, with eyes shut and unable to do much except snuggle up to its mum and drink milk, to being able to use its amazing agility and senses in order to survive by hunting (if it had to). Its mum could have already had another litter at this stage as she can quickly come back into season and pregnancy only lasts nine weeks. These kittens need to be potentially independent very quickly. If the kitten is not being cared for by people, the harsh world will be challenging, and cats need to be fast learners and become self-sufficient very early in life. We think of the age of six months being the equivalent to a ten-year-old human, a short step from puberty. This is a period when cats are very appealing but when their curiosity can also get them into trouble! Being aware of hazards in the home and outside can help to get your cat through this period safely.

We also need to give them the best start in health by arranging their vaccinations, worming, flea treatment and of course neutering (see Chapter 5). Getting kittens used to handling such as is needed for veterinary examinations, having their teeth cleaned and taking a tablet, grooming, going into a cat carrier and taking a car journey . . . making all of these things fun and not traumatic will have a huge effect on your cat going forward. Sarah Ellis's book *The Trainable Cat* and the International Cat Care website have lots of resources to help you get your kitten used to all of these human-world things. Getting the kitten used to its name and rewarding it for coming when called will help when you want to ask your cat to come to you. Providing a range of different foods which the kitten can be familiar with – wet, dry, etc. – can be very helpful in later life if you need to change diet. This is something cats can be very stubborn about, but the change may be necessary for the cat's health – for example changing from a dry to a wet diet because of kidney problems later in life. It is also a time to ensure you do not encourage behaviours in your kitten which will not sit well as it grows from small and cute to having larger and sharper claws and teeth. The best example of this is when we play with kittens with our hands and move them around, encouraging chasing, grabbing, biting and scratching. We wind them up when they are small and cute and then want to punish them to stop when the biting and holding with claws gets painful. The kitten then doesn't know what to do and this can damage the relationship.

The next stage is called junior and lasts from seven months up to and including the cat's second year of life and spans the time between childhood and mature adulthood. Think of a person between the ages of eleven and twenty-four – physically growing at a slower rate than a kitten but (hopefully) maturing and getting to grips with the world he or she lives in. The kitten becomes full size and knows how to navigate its life environment and to survive. It is certainly still learning about life, influenced by its earlier years in how it approaches things. The cat is in a curious phase of its life and this is the time you may be letting your cat outside, which is always quite nerve-racking! This is where your lessons on coming when called come into effect and you can let it out and then bring it back quite quickly, allowing it to get used to the ways in and out or learn how to use a cat flap.

As a cat reaches its third year, it has stopped growing and has reached behavioural maturity and transitioned from kittenhood fully into behaving like an adult cat. Between three and six years old is referred to as the prime period – the cat is fully grown up, healthy and active, sleek and shiny and navigating its life. It is the equivalent of twenty-five to forty in human years and our cat is indeed grown up. It is likely to know the habits and rhythms of the household and its territory.

Cats are referred to as mature between seven and ten years

old – it is equivalent to humans in their mid-forties to mid-fifties, physically and emotionally mature, hopefully fit and healthy and very much part of the family. They are likely to have us under control and have developed communication strategies of their own which are tuned into us. We can probably also predict how they will act in different situations and what they are going to do (see examples in Chapter 9).

Cats are classed as senior between eleven and fourteen years, which takes the cat up to the equivalent of about sixty to seventy-two human years old. Senior cats are real golden oldies and very precious to us. They become super senior when they are fifteen years and over, equivalent to over seventy-six in humans. When International Cat Care first used this ageing table, this stage was called 'geriatric', which probably reflected both human and veterinary language of the time. Geriatric is a word we don't like much because it conjures up images of illness or frailness. But cats, like people, can be very sprightly in their mid-seventies and upwards. So the name was changed to 'super senior', which is much more positive. Interestingly, people super seniors are cognitively and physically high-functioning individuals who have evaded major age-related chronic diseases into old age, representing the approximately top 1 per cent for human health span. I suspect more than 1 per cent of cats reach this older age – it is quite common now and probably the result of a combination of good nutrition, preventive healthcare and veterinary care, and a loving home. This is a time when we need to be super vigilant about watching our cats for signs of behaviour change in eating, grooming, weight, using the litter tray and interacting with us, which may alert us to health problems.

Whether your cat is neutered or not

Very few people are happy to live with an unneutered cat in their homes – male or female. The reproductive behaviour of cats is driven by a desperate need to find a mate. Female cats can become sexually mature from four months onwards, often at five or six months, depending on the time of year, temperature and day length, and are therefore capable of breeding and producing kittens themselves. Of course, this is not a conscious decision, but hormones activated by length of daylight drive behaviours which will have this outcome.

This means that they will be trying to let male cats know that they wish to find a mate. Likewise, male kittens will mature and grow up fast. Female cats will 'call' regularly, making strange miaowing sounds which many people often misunderstand as a cries of pain, and cats will be keen to get outside. If they do not become pregnant this behaviour will continue for about two weeks, stop and then repeat every three weeks or so until the day length changes again, but that might be at least six months later – a long time to live with a cat behaving in this way.

Having unneutered female cats in an area will attract unneutered entire males. Male cats may wander over large distances, spray urine, fight other males in the area and make caterwauling noises in an attempt to find a female ready to mate. Fighting males are also much more likely to spread and catch diseases and suffer from fight injuries such as abscesses. Because they wander over a large area, they are also at greater risk of getting run over on the road.

If you have not neutered your male pet cat, he is much more likely to roam over larger areas and may not even return. He may also spray inside the house, which gives off a very difficult smell to remove and is not pleasant to live with. Unneutered male cats may also be aggressive towards their owners because they are more reactive. Breeders of pedigree cats who keep male cats of a particular breed for mating mostly keep them in outdoor pens rather than in the house for these reasons, and to keep them safe as they are likely to wander and fight if they have free access outside.

Therefore, it is desirable to neuter kittens early enough (four months) to ensure that the above problems are prevented. Most people do not want to live with an unneutered male cat.

The cat's environment – your home and the area around it

Our cats' comfort and ease will be affected by us and our behaviour but also by where we live. Our home and the area around it will be its territory and a very important territory at that. If the cat lives totally indoors then a cat-friendly home is extra important, but even if it is lucky enough to be able to go outside, trying to 'think cat' for your home can only be a good thing.

We know that cats like to have choice and to feel that they are in control. When it is us who are actually in control of our cat's facilities (such as litter tray and type of litter, food, placing of bowls and beds) we need to be able to think this through with our cat mind, rather than what might be easiest for us. An example would be squashing all of the cat's plates and litter tray into a tiny space together. Or placing a litter tray in front of some large floor-to-ceiling glass windows or doors where other cats in the neighbourhood can sit and look in from the outside – this is very disconcerting for a cat and it may be put off eating or using the litter tray because it feels vulnerable.

Most of us are good at getting together all a cat might need but not necessarily thinking about where to put these things – because we are not putting ourselves in the mind of a cat. There are practical challenges to consider for both the facilities you choose and where you locate them. Thinking like a cat can help to develop a cat-friendly home for your cat. A relaxed and happy cat will be able to display its character with you rather than thinking about safety or threat. You want your home to feel safe, but also interesting and not boring. If cats are kept totally indoors, then providing a stimulating environment is very important as the cat needs an outlet for its quick brain and normal instincts, as outlined in Chapter 1.

Many of us are moving towards having houses with large open areas and minimal clutter, which may be aesthetically pleasing for us but may not be the most reassuring environment for our cats. Additionally, if you have more than one cat, having hiding places will be high on the cats' agenda. This is not necessarily because the cats don't get on well – even friendly cats may like to hide sometimes or want to avoid people or other animals – but if they don't, then having somewhere to tuck themselves away is vital. Additionally, when cats play, having somewhere to hide can ensure that play does not escalate into fighting as excitement winds them up (see Chapter 8 for more on this).

Cats like to rest where they feel secure and being able to watch what is going on below, so high places are valued. Resting places such as shelves or on the top of cupboards, or just going upstairs, are often preferred. You can now buy amazing cat contraptions that combine a scratching post with a tall combination of platforms at different levels where cats can rest – some reach right up to the ceiling. There are others which attach to the wall as a series of steps that lead to a platform higher up – but if you have more than one cat, make sure there are two ways up and down so one cat is not trapped up there by another. If you have a cat that likes to go on the top of cupboards, you can help by positioning other bits of furniture to aid getting up and down. If you have a height-seeking cat that climbs on shelves, then make sure you don't keep things there which will be broken if the cat suddenly jumps up and knocks them over.

If your cat likes to hide in small places, you can rearrange things to make sure there are spaces that suit – under the bed, in cupboards or behind furniture. Don't disturb your cat in its safe place unless you are worried that it's hiding because it is unwell.

Warmth, of course, is very important to cats, and yours may choose a place like the airing cupboard if it houses a tank of hot water, or perhaps in the boiler room. You can buy wonderfully cosy fleece blankets or use woollen blankets or duvets and you can even buy heating pads. Do cats use the beds we buy for them? Some do, especially if you choose a great position for them and make them very appealing. Many prefer to use our beds, of course, which have our scent and therefore feel safe, although they may well love a soft cat bed if you put it in the places they prefer.

They may use various places to rest during the day which follow the sun around the house to bask in the heat. So having a choice of beds in areas the cat likes will certainly be appreciated. Most of us almost automatically realise that cats don't like their beds to be on the floor and prefer higher positions. You will know your own cat and whether it likes to rest near people or away from them. Some will prefer beds with high sides or just an opening on one side; others won't mind being in the open.

The border between indoors and outdoors

The greatest challenge to the cat's perception of security in its home is the border between indoors and outdoors and what can breach that security. If you let your cat in and out by opening a door or window as it asks then it may feel very secure indoors, although getting back in quickly may be more of a challenge if it needs to wait for you to open up, especially if something is chasing it. If your cat only comes and goes when you are there, then actually opening a door or window yourself may make it feel more secure.

Many people have a cat flap, which makes life easier for both themselves and their cats. However, the flap could be seen as a vulnerable point in the defences of the cat's home, where any other cat could get in, and this may cause them anxiety (as they are constantly aware of this). As we know, anxiety can make cats behave differently as their minds are occupied by this rather than relaxing and interacting with us. If their food is near the cat flap, then clever cats outdoors may come in during the night or day and help themselves to it. Your cat will know this is happening even if you don't and you may notice a nervous cat glancing at the cat flap as you put food in the dish. Nowadays you can purchase cat flaps that open in response to your cat's microchip, ensuring other cats cannot get in.

Cats love to look out of the window to see what's going on and, if they are upstairs, can get a great view of the territory.

However, they do seem to be somewhat confused by very large floor-to-ceiling windows where it is hard to tell the difference between what is indoors and what is outside – a small window with edges and a nice windowsill is quite reassuring! As well, floor-to-ceiling windows or doors do not give cats anywhere to hide if they see things outside which seem challenging and this may make them feel vulnerable – coming face to face with another cat or a dog which is in the garden and the other side of the glass may be hard to compute when your instincts are telling you this is dangerous.

Watch your cat – does it seem to be unsure near full-length windows or doors? Is it looking out quite fearfully? Does it have behaviours that seem to suggest it is anxious? It may be worth putting some form of frosted cover on the bottom part of the glass (you can get a kind you don't need to glue but where sticking relies on static). Give the cat a high place to view through the glass so at least it's not on the same level as something outdoors. Put furniture or plants in front of the window that the cat can hide behind.

Presentation of food and water

This brings us on to how food is presented for cats. Cats don't particularly like their whiskers to be touched, so providing dishes which are wide enough for them to eat from without touching the sides may be welcome. But food falls off a very flat dish (and makes a lot of mess for you to clear up), so a wide one with some type of edge to keep the food in place and provide something for the cat to push against as it eats may be the best solution. If you have a very flat-faced breed of cat such as a Persian or Exotic it is likely that its skull shape has affected its jaw and it may find picking up food difficult, so you may need to experiment to find the food and bowl shape which suits it best. Ceramic or glass food bowls are probably better than plastic, which scratches easily and can give off a slight odour that your cat may not like.

If you talk to people who are experts on cat behaviour you will find they are very keen on making cats work for their food, especially those kept totally indoors and therefore unable to use their minds and bodies for hunting. Cats that have to hunt for their own food would naturally spend up to six hours a day foraging, stalking, catching and eating the small prey they need. They need about ten mice-sized meals a day and considering only about a third of hunts actually result in capture, you can see how much time and energy this takes. This is what cats are designed for. Behaviourists would argue that having food twice a day in a food bowl in the kitchen does not replicate this at all and the cat does not get the mind or body stimulation hunting would naturally bring. You can make or buy toys and equipment that make cats work/play to get food to encourage them to actively seek it – there are lots to be found on the internet under the titles of 'interactive' or 'puzzle cat' feeders. More advice can be found on the International Cat Care website too.

We all know that we need to provide our cats with fresh water and the obvious thing is to put it in a bowl next to the food. However, when cats have to look for their food outdoors, they probably eat and drink separately. Having water near the food may deter some cats from drinking enough (cats on dry diets need to drink much more than those on tinned or pouched diets, which contain a lot of water). Behaviour experts advise putting the water separately to the food and having more than one bowl if you have several cats.

The experts definitely don't suggest putting food and water in one of those cat-feeding stations that have two bowls next to each other. Cats don't seem to like food next to water and often in these situations some of the food falls into the water and contaminates it. Also, because cats have a much better sense of smell than us they may not like the smell of tap water and may prefer it boiled. Some cats love those drinking fountains which have constantly running water and the same guidelines apply to bowls as for food. Keeping the water clean is important and ensuring it is topped up so the cat can reach it easily may be appreciated too.

Providing toilet facilities

Once again, it is not just providing a place for your cat to be able to toilet that is important – it is the choice of litter, tray, positioning and cleaning. Obviously, if your cat does not go outside, then a litter tray is an essential. Some people provide a tray even if the cat does go outside, if they shut the cat in at night or the cat is very young, or old and needs a bit of support.

A basic litter tray is simply a large open tray that is big enough to turn around in (suggested length one and half times the length of the cat, not including the tail) and which can hold enough litter for a cat to use and scratch up to cover urine or faeces without scattering it all over the floor. Kittens or older cats with arthritis may need lower sides to enable them get in and out of the tray easily. If you want to add a bit more sophistication you might choose a tray with a lid with the aim of providing some privacy and containing smells. Cats vary in their preferences – some like the privacy, others may not like the confinement. It also matters what is happening outside the tray – are there other animals or people who are making the cat feel vulnerable? The same applies for the position – is it in a busy part of the house or in front of a window or is it positioned somewhere a bit more private? Don't line up the litter tray with the food and water – separate all three.

The same applies to the different types of litter which are available – your cat may be happy to stick with the one it used as a kitten, but you may want to choose something different. Take the time to mix towards change so the cat gets used to it as well. Cats are not thought to like polythene litter tray liners, which can get caught in claws, or deodorants that are more for people than cats. Whatever you choose, keep the tray clean as this will be important to your cat. You may want to have a litter tray in more than one place depending on your house and your cat, and you definitely need multiple trays if you have multiple cats.

Somewhere to scratch

Cats need to scratch to maintain their claws and mark their territory. If provisions are not made for this, then they will scratch items of furniture. Scratching posts should be as tall as possible to allow your cat to scratch vertically at full stretch and to be stable. Panels can be attached to walls at the appropriate height if space is at a premium. Some cats prefer to scratch horizontal surfaces, so a variety of scratching areas should be provided.

Having other cats in the household

One of the major choices we make is how many cats we have, as our pets or that we encourage from outside into the house. This may be because we love cats and we want them all to love each other, but unfortunately, especially for cats, this may bring tension to our homes. Some cats enjoy sharing their homes with other cats and form bonds with them; for others this can mean anxiety. It is something to consider carefully when we feel we want more cats. We need to consider our homes carefully, looking at how much space we have – are there places for cats to each have their own space? Have we arranged all the things they need so there isn't competition between them? Bear in mind that cats have a strong instinct for defending territory and we can push otherwise relaxed pets into a conflict situation if we introduce new cats. If we do decide to add cats to our household, we need to do so carefully and slowly (see International Cat Care website). It can be done successfully but requires patience and depends very much upon the individual cats.

Other cats in the area

Having other cats living nearby, in the garden or even coming into the house can affect how relaxed or stressed cats feel and this can impact on their interaction with people. I am reminded of a cat we had many years ago which appeared as a rather feisty grey and pale orange tortoiseshell kitten; we took it into our house about a year before we got two female Siamese kittens. That cat, called Peru, was somewhat of a terroriser of the new kittens but, as they got bigger and older, they started to retaliate somewhat. Peru then decided she would move to the cottage down the road because her life had become rather risky once the Siamese cats had learned to get in first before she bullied them. We continued to feed Peru, so she would come back for food but would relax at the other cottage. Interestingly, if we saw her there she would be ultra-friendly with us, purring and rolling around. If we tried to stroke her when she came to the house to eat, and was therefore in danger of meeting the Siamese cats, she would be spiky and very unapproachable. You would have thought it was two different cats.

I read something years ago, a theory about a cat's 'aggressive field'. When cats are relaxed and not feeling threatened or stressed, this aggressive field is smaller than the cat and we can touch and stroke it without a problem. When the cat is feeling under pressure, the field widens until it is several metres around the cat. Any intrusion into this area risks attack and will certainly elicit threatening behaviour. Presumably this is similar to what is called the 'flight zone' of an animal, the area around it that, if breached by something threatening, will cause the animal to act to escape. The size of this space depends on the willingness of an animal to take risks and increases as the perceived risk to the cat increases. It's a great example of how different behaviour can occur within the same cat with the same person but within a different environment.

Cat-friendly garden

It is not just the indoor environment that can be made cat-friendly. As we discussed for indoor spaces, cats also like to have somewhere to hide in the garden, so vast areas without plants or other places to hide may not be favoured by them. Cats that are nervous outside may not feel safe without somewhere to hide. This gives the keen gardener the excuse to plant hedges, position pots or develop sitting or even artistic areas that cats can use to navigate and explore or as shelter. Remember they like to climb and have high views or walkways and to sunbathe too! Many people now fence in their gardens so their cats are safe from going onto the road and can enjoy fresh air and all the interesting things which happen in a garden – insects, small animals, wind-blown leaves, etc.

Preserving cat scents

We are familiar with our own homes and with their layout and where things are – this is mostly based on our sense of vision. However, as we know, cats are highly sensitive to scents and have a great sense of smell – think of them having a scent profile of our homes as strong as our visual perception of them. We know how much effort they put into marking their scent around and how they are affected by strange smells, especially if they are nervous or anxious individuals who find it difficult to deal with change. Having familiar scents in all the right places will make our cats feel safe and relaxed. Being aware of how much we use strong-smelling chemicals or scented products in the house and of not washing their beds or favourite sitting places too often will ensure that some safe scent areas remain for them. Product advertising encourages people to have ultra-clean and fresh-smelling homes, but this may not be cat-friendly and may be a good excuse not to follow these advertisements too closely. A relaxed cat will hopefully be a much more interactive cat.

The choices we make for our cats

Indoor/outdoor lifestyle

Living indoors without access to the outdoors deprives a cat of the experiences of an outdoor lifestyle. There are of course many reasons for keeping cats indoors, including the dangers of outdoors or not having an outdoor space at all, or owners keeping their cats in as they fear for their pet's safety. Some indoor cats may not adapt to this and may suffer from a number of physical or emotional problems which may have been brought on by boredom, frustration or lack of an outlet for activity and opportunities for using those amazing senses that drive them. Cats may then sleep, groom or eat more to compensate for the lack of stimulation. Indoor cats can develop health problems associated with a sedentary lifestyle, for example urinary tract disease, over-grooming issues and eating disorders. Owners can help by fencing in all or part of the garden or even a patio safely to give the cat some outdoor access if possible.

Predictability

Another way we influence our cats is in how predictable we make their lives. As with people, some love predictability and others loath it. But we have discussed cats as liking to have control and to know what is happening, so being aware that something is going to happen because there is a pattern to the day and their activities may help some cats, especially those a bit more nervous about life. Our cats often teach us what they want and use communication methods to be able to control us to help them.

Physical health and mental wellbeing

One of the biggest influences we can have on the quality of cats' lives is to protect them from disease and to act to help them if they become ill. We have only had ways to protect our cats from disease by vaccination for about fifty years and before that we had little information about the diseases and problems affecting their heath. How we keep our cats healthy is similar to keeping ourselves healthy – feed well, give whatever preventive care we can to protect them, be willing to act quickly if they are unwell, and work with a good cat vet in a cat-friendly veterinary clinic. Cats do challenge us with their health because they are very good at hiding signs of illness or pain (more on this in Chapter 5), so we need to be aware of the signs that they do give. We know that how cats are feeling, whether they are suffering from long-term stress or anxiety can affect their health, and health can affect wellbeing, so being able to understand what your cat is feeling is important. It can have a big effect on our relationship with our cats if they are quiet, hiding away or unhappy to be touched because they feel unwell.

How we approach cats

We love to love our cats and often want to cuddle them. Some cats like this, some tolerate it, others will not embrace it at all. As always, cats are individual in how much they do and do not want.

Traditionally women have been more involved with caring for cats than men. However, when researchers looked at cats and how they approached people they did not know, they found that the cats approached boys, girls, men and women similarly. It was the people's reactions that were different – the men interacted with the cats from a sitting position and the women, boys and girls got down to the cat's level. And, as you can probably guess, the children couldn't hold back from approaching the cats and the boys were more active and even followed the cats as they moved away from being pursued, while the adults waited for the cats to make the first move. The women interacted with the cats from further away and talked to them as they approached, and stroked them more when they came close enough; the cats reacted to the encouragement and were happy to interact. Another study showed that if a person responds to their cat when it wants to interact with them, the cat in turn will comply with the person when they want to make contact – a mutually responsive and positive relationship. Presumably the initial requirement is a cat that is happy to interact – what we would call a friendly cat. What else do we know from research? If cats live with a family or a lot of people, they give each one less attention than they do if they live with one individual – that seems to make sense. In the same way, single cats living in a home spend more time interacting with people than cats living in a home with multiple cats. That is common sense, too. Owners who had just one cat were more tolerant of its foibles than owners with multiple cats. That is understandable too – not only are there more cats more to care for, but multiple cats often have issues between themselves, so people have even more to deal with!

It is when something extra happens, such as another cat coming into its territory, other threats or changes or being in pain that causes a cat to act in a different way, and it is up to those who care for the cat to turn detective and try to figure out what is causing that change. Cats may behave in a way that people may refer to as 'problem behaviour' – more realistically described as the cat acting in a way that is natural to it and trying to cope with the threat or stress it is experiencing. The problem part is that it is happening in a way or place which people do not like. Examples may be urinating in the house or acting in a way to push people away, which may be interpreted as aggression.

My friend Vicky Halls, who I mentioned earlier has written best-selling books *Cat Confidential*, *Cat Detective* and

Cat Counsellor, can lead you through some great examples and fascinating stories about cases she has covered when she was a cat behaviour counsellor for over twenty years. She now works for International Cat Care looking at how we provide the best solutions for unowned cats depending on their lifestyle, and how best to care for cats in homing centres by understanding their individual needs, and what homes would suit them. Discussions with Vicky over the years have shaped the way I think about cats and what we consider to be good cat welfare – we have had many good discussions, challenging our thoughts with the 'But why?' question and looking at how research can help us to develop practical ways of understanding and helping cats. Even after so long, we both realise there is so much we still don't know about cats.

But once you have your cat, it is neutered and your home is secure and routines are in place, you will be able to ascertain how your cat is likely to approach most things. This will also be affected by how you interact with your cat – whether you pursue it to cuddle it or whether to allow the cat to do the approaching and let it show you what it needs.

3.

Your cat's personality

PEOPLE LOVE TO pitch cats against dogs and ask which is more intelligent. Are cats intelligent? Are they more intelligent than dogs because of their ability to live independently if necessary or should they be judged to be less intelligent because they don't do what we ask them to, they are not obedient? Is intelligence the sum total of intellectual skill or knowledge or is it the ability to learn new things or distinguish one thing from another? Perhaps it is more the ability to adapt to changing circumstances. If intelligence is measured by adaptability, then the cat is top of the class. As a species cats can survive in almost every type of environment – desert, jungle or even in very cold climates.

They are highly versatile and seem to learn quickly from their experiences. Kittens have to be very fast learners because, as we've seen, maternal care is relatively short – they have to go from birth, when they are unable to see or hear, to being able to tackle and catch prey within a few months, by which time the mother cat is pregnant again and producing more kittens. Kittens need to be curious and learn by watching and experiencing (helped by their mother), while remaining wary of dangers. They may seem vulnerable and soft, but the survivors are tough and fast to adapt. And while there is help for them when they are young, they will grow to the life of a solitary hunter, self-reliant for food and for survival. Of course, in the wild, unfortunately many do not survive the challenges they face.

Motivations to behave – a glimpse into cat emotions

Cats are emotional creatures. Emotions are the 'feelings' that motivate the cat to react in certain ways to overcome the challenges of survival and they are vital to learning and reacting to everyday life challenges.

Cats are not pack or herd animals and do not need to have other cats around for their wellbeing (except of course when they are kittens). However, they can be what is called 'socially flexible' in that they can live in groups if the circumstances are right and can form relationships with other cats and other animals, something we can observe when we have them in our homes.

So far, we have seen how cats sense the world around them

and the behaviour available to them to deal with what life challenges them with, hunting to survive, communicating to reproduce, defending territory – all to do with survival.

Cats use the senses we have outlined to perceive the world they live in and take that information and decide what to do with it – sometimes it is a reflex action which does not require thought, and sometimes the cat considers and reacts, learning from the experience. How it reacts will depend on many different things, but its emotions will motivate it. Unlike people, cats don't think about, meditate on, evaluate and give serious thought to their behaviours, thoughts, attitudes, motivations and desires, but their emotions affect how they act and change according to the circumstances.

An emotion is a feeling such as happiness, love, fear, anger, frustration or hatred, which can be caused by the situation we are in or the people we are with. We may think of emotions as linked to crying or laughing, but emotions are also about being attracted to doing things we like, or to things which reward us in some way. This can be as basic as looking for food, water or warmth. So, while we may not consider that our pets have emotions, they do in fact experience some of the same ones as us and understanding this can help us to consider how our cats are feeling and what is making them act in a certain way. What we must not do is align cats too closely to ourselves – if we simply think about them in an anthropomorphic way, we may completely misinterpret what is going on.

For cats, hunting is something with a strong drive and associated with the cat's curiosity; undertaking the behaviour probably gives a feeling of reward or pleasure. There are strong emotions, too, involved with the motivations for reproduction so that cats can behave in a way to attract a mate and interact with that mate in order to bring the next generation of cats into the world.

If a female cat has kittens, then it has a strong motivation to look after them. Caring for kittens and sometimes other cats or different animals (perhaps grooming them) is seen as a reaction to an emotion. If kittens become separated from their mother, they may experience the emotion of panic and will make sounds which express their distress – these sounds in turn will motivate the mother to look for them and bring them back to safety.

Playing with their siblings allows kittens to learn how to relate to and act with other cats, as well as honing their hunting behaviour. Mature cats don't have to interact to survive and stay well in the way that a social species like dogs or humans do, but they may still enjoy the experience of playing.

We can all relate to feeling anxious or fearful about something which is threatening us or that we think is dangerous – it can range from a fast jolt to the system to a lower level of feeling that we might call anxiety or dread, and will motivate the cat to avoid threat and harm. The experience of pain is included in the emotional category of fear because it protects the body if it is damaged and helps to prevent further damage. We also know that chronic pain can lead to feelings of depression and removes the desire to do other things.

It is interesting that frustration is considered an emotion, but when you think about it, it is a strong motivator to act to overcome something which is blocking what the cat wants to do or to get to. Frustration motivates the cat to try harder and is associated with choice – being able to get what it desires. It can also be experienced if the cat wants to get away from something it fears but is prevented from doing so. All these emotions may occur together and maybe in opposition to one another – such as wanting food but being frightened to go to it because there seems to be something perceived as dangerous in the way.

What is personality?

We have seen how cats can make good pets and how their early life is so important. But what about personality? Personality refers to individual differences in patterns of thinking, feeling and behaving. Anyone who has owned more than one cat will be very aware that each of those cats has a very different personality, able to explain how they differ from each other and what their habits and preferences are. The words 'individuality', 'personality' and 'temperament' are all used to explain the particular ways a cat behaves that make it different to others and give it its own 'style'. You can probably figure out that researching personality is not easy

– those doing the research have to try and define different factors which make up a 'personality', such as friendliness, curiosity, nervousness, excitability or boldness, being vocal, being active, interactive, etc., and then bring them all together to give the cat a personality.

Boldness is not a word we use very often in relation to people, but it does come in useful in describing cats. Defined as a 'willingness to take risks and act innovatively; confidence or courage', for our cats it may be the difference between avoiding something novel or challenging and investigating it. Interestingly, some cats, even though they have not had a good period of interaction in their early months, can form some type of a relationship with people. This trait can help the cat to make contact with people even if they are unfamiliar with them but is probably not quite the same as 'friendliness'. These cats may be the bold ones that overcome fear because they are motivated to partake of the benefits of being with people – they may not necessarily want to interact with many but find a way to be with and communicate with an owner who is open to their needs. If you combine boldness with confidence generated by a good start in life and friendly genes, then you have one of those extraordinary cats that wants to be involved with everything that is going on and has a strong, positive character around people.

Research into personality is hard enough to undertake with people, where we can ask them directly for their preferences, likes and dislikes and delve into their thoughts. Of course, when people do research on cats, the presence of the researcher can affect the cat's actions, so they have to find clever ways to compensate for this. The same can be said for us owners – we will have an effect on our cat's actions and behaviours, too, and that will in turn affect how we think of our cats. And while temperament or personality is thought to be quite 'set' in that a cat will react in a similar way in different circumstances, how it deals with a situation may be affected by age, health and the environment it is in. An example may be that the cat is confident with you but nervous around strangers. What is clear is that a cat's personality is formed by a complicated interaction between genetics and experiences, especially those of its early life.

The individuality of cats

The individual personality of cats has been recognised in science from the 1960s. Recently, as I mentioned earlier, I met Dr Dennis Turner, one of the early researchers and writers on cats and their behaviour – his book, with Patrick Bateson, is called *The Domestic Cat: The Biology of its Behaviour*. During his lecture he mentioned in passing that 'cats are the most individual of all animals' and that caught my attention. I asked him to elaborate – did he mean that cats are more individual in their behaviour than other animals? He explained that when you study cat behaviour you have to take into account and correct results for individual differences. He was unaware of studies into other species having to do this. We cat lovers love to hear that our feelings about our cats being real individuals have scientific backing!

The effects of breeding/genetics

If we are thinking about genetics and how they influence behaviour, then it is worth looking at pedigree cats to see if that can give us a clue. Are there clear differences between the behaviour of pedigree cats and moggies? And are there differences in behaviour between the different breeds of cat as there are in dogs? Over years of association with man, dogs have been bred for certain specialised tasks such as guarding, shepherding, tracking or even fighting. They were expected to perform those tasks and would probably not be bred from if they did not do them well or as expected. Others have been selected more as companions and usually have very friendly, people-orientated characters.

Of course, dogs are well-suited to living in human society because they have a similar social setup and they fit into our 'pack', or we fit into theirs. Cats, however, having developed from solitary hunters and without a hierarchical structure even if they live in more social groups do not have the same inherent cooperative natures. They have lived with man for thousands of years, but usually by their own rules and without any expectation to perform particular tasks. The cat of old, which lived by its wits around the homestead and was lucky to get the occasional drop of milk or scrap of food, if it was selected for anything at all it was for its hunting abilities, to keep vermin from stored food and later for its rodent control capabilities. So, most cats around the world breed randomly with a mate nearby – they are remarkably consistent in their size and maintain the ability to look after themselves, staying agile and motivated to hunt. In terms of natural selection, kittens of good hunting mothers would be most sought after in such circumstances and may have had some support in terms of being given scraps around the homestead. Or perhaps not fed in case that interfered with hunting motivation.

We talk about cat 'domestication' but many feel that cats are not really domesticated. A domesticated animal is one that has been tamed and selectively bred to be kept by people for working, or for food or as a pet, and is different to its wild ancestors. Our pedigree cats have been bred from chosen pairings of animals in the same way we work with our truly domesticated animals such as cattle and sheep, which have mates that are selected to develop or perpetuate traits that people want. A pedigree cat is one that is bred in a very controlled way, with males and females chosen from a small number of similar cats. This leads to a certain 'look' that influences body size and shape, coat length and colour (including lack of coat) – and perhaps behaviour too? Certainly, by being able to choose which cats produce kittens, cat breeders can select from cats that are friendly towards humans and there is no doubt that many pedigree cats can be more people-centred than their non-pedigree counterparts. We can maintain this trait by continuing to choose such cats through the generations, thus influencing their genetics. By doing this, and by raising kittens so that they get optimum interaction with people and experience of the human home, we can really give kittens the best start to becoming pets that will not find being with people stressful.

However, interestingly, some of the breeds seem to have almost become too people-orientated and suffer from something which our non-pedigree cats don't usually – separation anxiety. Independence is a recognised cat characteristic across the world, perhaps because our non-pedigree cats or moggies are often the result of a pet female cat able to go outside while in season and being found by an unneutered tom cat. This tom may be living on its wits (and not being caught to be neutered) and so is what we might call 'reactive' in that it has to be alert to danger and bold enough to find a mate.

Our non-pedigree cats are certainly more friendly than their wild ancestors but still retain a great deal of independence and are likely to be able to live without people if they have to, reverting to natural hunting and survival behaviours. Part of me likes to think that most cats are not truly domesticated, but my mind was slightly changed when I flew from Moscow to Prague for cat veterinary conferences. In the plane opposite me was a lady with a zip-up bag-type cat carrier, which of course immediately caught my attention. I thought about what my cats would think about being on a plane – not amused and pretty scared, I thought. There were some quiet miaows from the carrier, and I watched as she began to unzip the bag while considering that any of my cats would make a bid for escape and find themselves some small and inaccessible place on the plane to hide and likely cause some major incident. The cat that emerged was a Sphynx – an almost hairless type of cat. She put a rug on her lap and lifted the cat onto it then put another little blanket over the cat, which curled up and seemed to go to sleep – it stayed still but I couldn't see if that was due to comfort or fear, but it didn't try to run away or hide. I was thinking about litter trays, etc. when I saw her stroke the cat under its tail and seem to catch urine in a tissue – had she trained the cat to urinate in such circumstances? It was all done very quietly, and nobody would have noticed except that I of course was fascinated! When we got to our destination, she put the cat back in the bag and got off the plane. My immediate thoughts were that perhaps some cats were now indeed 'domesticated'! Did I like that? No, not really. I prefer my cats more natural, less compliant and independent. However, the question it raises is whether that cat was actually as relaxed and not stressed by the activities of its owner and the circumstances she put it in – if it was happy, then that is a good thing.

Many breed lists will wax lyrical about the different

personalities of the different breeds. However, personality relies not just on genes but on early experiences in the kitten's life, and for most cats the variation between individuals can be as great as the so-called different personalities of the breeds.

There are also some recognised breed dispositions, such as the Siamese being perhaps a talkative or noisy breed and Persians seen as quiet and less active. Of course some breeds are entirely dependent on their owners for care – for example, the long-haired Persian which cannot keep its coat (with its extra-thick undercoat) free from tangles, and at the opposite extreme, the Sphynx (which is virtually hairless) that needs bathing and cleaning to deal with the oil which should coat the hair but instead remains on the skin; it also needs protection from sunburn and cannot cope with cold temperatures.

What we do know is that the differences between individual behaviours within breeds are generally greater than breed differences – once again the individualism of the cat comes to the fore.

What has cloning taught us?

In the early 2000s there was a breakthrough in cloning animals (copying genes from one animal to form a second, genetically nearly identical to the first). Dolly the sheep was the first successful example. A couple of cats were also cloned. CC, short for Copycat or Carbon Copy, was the result of taking a cell from a tortoiseshell and white (known as calico in the US) cat called Rainbow. Single kitten CC, however, was tabby and white and did not look the same as her generic donor because coat markings result partly from random events during development. And, of course, looking at how cat personalities develop, those first two months of life are crucial for any individual – if they are different for different cats, will the personality be the same? One report says that CC and her genetic parent had different personalities in that CC was shy and timid, while her host was playful and curious. You can imagine the early weeks of CC's life were probably filled with lots of scientists interested in her health and development – perhaps very different to a kitten being born in a quiet pet home.

While many different animals have been cloned for different reasons, a couple of cats have been cloned because their owners wanted to replicate their pets which had died. Looking at these, the money you would spend may not be worth it – the cat may not look the same or have the same personality. But do we really want a cat that is exactly the same as the previous one? Most of us enjoy our cats for their individuality, the things they do or the way they do them that are unique to that cat and, if recounted, would allow us to identify it among others. Perhaps it's better to acknowledge that a cat is a one-off and special, and move on to a new and different one, appreciating it for its own personality and individuality.

Does coat colour affect personality?

There has also been some investigation into whether coat colour has an effect on behaviour, and several thoughts have come from this over the years, as well as quite a bit of folklore! There are lots of anecdotal stories about coat colour – mostly attributed to tortoiseshell cats and ginger cats. Tortoiseshell cats have a reputation for 'bad behaviour', perhaps being less tolerant of handling by people when they are not in the mood, and intolerance of other cats and of other cats in their territory. Ginger cats are believed to be more friendly than others, black cats thought to be more tolerant of crowding and urban life, while white cats are said to be shy and quiet. Research continues.

Coat colour is a very interesting area though. Genes controlling coat colour are located on the X chromosome. Remember that male cats have X and Y chromosomes, while females have two X chromosomes. The genes have two alleles, ginger and black, that are mutually exclusive, which means they cannot occur together on the same gene. Offspring get one from their mother and one from the father, which then combine to produce the eventual colour of the kitten. A female cat (with two X chromosomes) carrying one copy of a ginger allele on one gene and one of the black alleles on the other will turn out to be tortoiseshell. The black gene is more common in the cat population, so a male cat which can only carry colour on its X chromosome will be whatever colour that allele leads to, ginger or black. Because being ginger needs either two ginger alleles in female cats or just one on the X chromosome of male cats, ginger male cats are more common than ginger female cats. For the same reason, a male cat inheriting genes normally cannot be tortoiseshell. Occasionally male tortoiseshell cats are found, but they have inherited unusual genes patterns, such as XXY, and there is probably less than a 1 per cent chance of this occurring. Male tortoiseshell cats are also usually infertile.

But of course, there are more cat colours than just ginger, black and tortoiseshell – how does that come about? Well, there are things called 'dilution genes' that act to create a lighter coat colour and so create grey and blue coats. Similarly, the orange can be diluted to cream or yellower colours.

How then do you get a white cat? White cats have a special gene called a 'masking gene' which overcomes other genes and is dominant to other genes, so the cat will be white. There are also genes that create spots of white which can produce the white paws or chest or blobs of white that we like so much on a coat. Coat patterns are controlled by other genes and lead to tabby patterns or agouti colouring (where the hair has striped bands which lead to what we might think of as a natural coat colour as seen in rabbits –and we see in, for example, Abyssinian cats). Having a white coat is also closely linked to deafness if the cat's eyes are blue, so this of course could cause behavioural differences to non-deaf cats.

How then do we get the colouring we see in Siamese and

other breeds, where their extremities or points are a different colour to the rest of the coat? This is caused by a change to the gene that creates colour because it is sensitive to temperature – thus extremes of the body such as the ears, face, paws and tail are cooler than the main part of the body and the hair which grows there is a different colour. You can also see this when a Siamese cat has an operation that requires its hair to be shaved off. It may grow back in the first instance rather darker because that area of the skin is cooler. As that hair is then replaced after the shaved part has grown back, it will come back lighter.

There have been a few studies on coat colour and personality. One online study questioned owners about all sorts of things about their cats to ensure they did not realise the questions were about colour and how the cats behaved. According to the replies, tortoiseshells, black and white, grey cats and white cats were more frequently aggressive towards people. Female cats were reported to be more aggressive than male cats when being handled. Looking a little more closely shows that actually results showing 'aggression' were low and may have been influenced by owner expectations because they had possibly already heard about myths such as torties being more aggressive. Remember too that behaviourists don't like the use of the word 'aggressive', which comes across as 'attacking without provocation'.

It is interesting to see that female cats came out with higher scores, but this of course may just mean that they are more fearful and don't like being handled. Interestingly, a Spanish study found that bite injuries mostly came from Siamese cats, but look at this more closely too – people choosing a pedigree such as a Siamese may want a more intense relationship with their cat and may expect to be able to handle or cuddle it more, which may result in what we have classed in our diagram (see page 108) as 'protective behaviours', trying to push people away. An Australian study did not show any link between cat biting, breed, sex or coat pattern.

Another study of cat owners in Mexico found that grey cats scored highly for being shy, aloof and intolerant, while orange or ginger cats came across as being trainable, friendly and calm. Tabbies scored highest on being bold and active, while tricolour cats came out as more stubborn.

Yet more research has asked people to rate the personalities of cats based on their tendencies to be active, aloof, bold, calm, friendly, intolerant, shy, stubborn, tolerant and trainable. Results concluded that ginger or orange cats were friendly, while black cats and white cats were regarded as more antisocial. White cats came out as more shy, lazy and calm, while tortoiseshell cats were more likely to be depicted as both more intolerant and more trainable. Black cats came out as having less extreme character traits. It's true that even more research is needed in this area and as we have seen, there are many influences on cat characters so just judging a cat by its colour may not be helpful in choosing a cat to suit us.

There are several theories about how coat colour could

influence personality – one that coat colour pigments (melanin) are produced by the same biochemical pathways and chemicals which play a role in brain activity, another that genes for coat colour are on chromosomes close to other genes that influence the nervous system (and thus behaviour).

People seem to prefer light-coloured cats and homing centres report that it takes longer to home a black cat, perhaps because they are difficult to photograph well and may not stand out from others to give them 'character'. Gingers usually home very quickly – perhaps because of a mixture of an attractive coat colour and a reputation for being friendly.

Experiences and personality

We saw in Chapter 2 that the early experiences of kittens are vital to their future development. Kittens handled regularly during the first two months of life seem to be more confident approaching objects or situations they have not come across before. The timing and amount of early handling and number of handlers a kitten has influences later friendliness toward people, but all of a cat's experiences will shape how they react to challenges and learn more throughout life.

There are also two other factors which can alter a cat's behaviour – we think of personality and temperament as being 'fixed', but substantial traumatic events in the cat's life, such as accidents, disease or injury, can affect them profoundly and in ways that might alter their behaviour permanently. An encounter with a dog that causes injury may mean that the cat will be not only fearful of all dogs but may become afraid of many things it previously ignored. The encounter may bring on a complete loss of confidence in its ability to escape and any subsequent threat will cause the cat to overreact to the situation. We see the same effect in people who have been attacked and their lives affected thereafter. Confidence may return slowly.

Intense nursing of ill cats has also been reported to bring about changes in character. Some previously 'untamed' cats have become friendly after being intensively nursed for weeks. Behaviourists explain that their having to accept human attention actually 'floods' or overwhelms the cat because it cannot escape and so has to accept the situation. Flooding is not viewed as a healthy act and not something to do purposefully to animals. Whether during this flooding the cat is forced to realise that the person is not doing bad things to it (which it had totally avoided experiencing during its previous life and had successfully survived) is not known. It is also not reported whether the cat is then friendly to people in general or just to the person who cared for it – or indeed whether it is something like Stockholm syndrome, where in a kidnapping or hostage-taking situation feelings of trust or affection are felt by a victim towards their captor. Alternatively, the cat may have been living a more street lifestyle and living on its wits but had actually learned about people early on and so during the sensitive period had the basis of a relationship to fall back on.

There will always be cats that buck the trend or don't behave as we expect, but it is worth trying to understand so that we can work and care for these cats better and give them a chance, or if they are truly traumatised by our presence, to accept this and find alternative lifestyles for them.

So, to summarise, the unique personality of your cat in how friendly it is towards you arises because of its genetics and experiences with its littermates, how positive its learning was in the environment it was kept in, how its mother cared for it and reacted to people within sight of the kittens, the quality of its interaction with people and, of course, the current situation it finds itself in and whether it is threatening or not.

4.

What do we want our cats to do for us?

AT THIS POINT we have all the background we need for a basic understanding of cats. There are numerous and complex reasons why we like to have cats in our lives, some of which are very beneficial to cats and some that may be more about us than the cats. Where it is more about our needs than those of our cats it might be worth an honest consideration of the cat's position and ways to improve its welfare. Reading through this chapter again, I see some of it has been an outlet for some things I feel rather passionate or concerned about – where unfortunately human need/want in some cases definitely trumps cat welfare.

Why do we love cats?

Cat/person relationships are all individual – the quirks of the person and the quirks of the cat combined! And just like person-to-person relationships, they change over time and we come to love (or at least accept) some behaviours – and maybe grow more annoyed by others! There is no doubt that we have in our minds how we would like our ideal relationship with our cat to be, even those of us who know we must accept people or cats for the individuals they are.

For most people the 'ideal' is likely to be living with a very loving and cuddly cat that is happy to be around family and friends, is not 'destructive' or 'dirty' in our homes, is not 'aggressive' or 'bad-tempered', returns our love and provides companionship. For others it may be that they want a cat to fit in with their own beliefs, such as not eating meat or catching other animals. Of course, cats that do not have access to the outdoors cannot catch prey, but there are other pressures on cats that do not have an outlet for innate hunting behaviours and owners must try and compensate for this – there is no simple answer! Most of us these days would prefer that our cats do not catch wildlife and bring home 'presents', but that unfortunately is denying the very essence of the cat. Some people are happy to have a more distant way of living with a cat in the way we would enjoy feeding and helping a wild animal and, for some cats, this is an ideal scenario avoiding great anxiety – the benefits of food and shelter without having to be too close to a person. Other people need a closer relationship and reassurance from the cat that they are important to it (which is quite difficult for cats to do!). The perfect match occurs when cat and human desire the same thing.

We seem to have a cat-sized hole in our lives which we want to fill, but why? Is it to care for cats, to save them from a worse fate, for companionship, for power or for creativity? We don't do anything without reward a cynic might say, but the reward may simply be the satisfaction of giving or sharing. Of course, just as cats are individuals so are people, and our motivations vary considerably.

In a study undertaken in the UK in which almost two-thirds of those who took part were female, different types of relationship were explored using a questionnaire to gather information about different emotional elements that could underpin the relationship. Different types of cat owner relationship were noticed that seem to explain the type of relationships people had, from 'open' to 'remote', 'casual' to 'co-dependent' or 'friendship'.

Because it is thought that how we feel about something may be affected by our interactions, the study looked at the owner's emotional investment in the cat, the cat's acceptance of others, the cat's need to be near to its owner and the cat's 'aloofness'. Results from a questionnaire reveal what owners think the cat is and does and in turn how this affects what owners feel about their cats. For example, owners may believe their cat is loyal or faithful or emotionally close to them and that it really enjoys stroking or interacting. When you look at cats and their behaviour, this becomes more complicated as people may feel that their relationship is close because the cat's behaviours are what the owner needs. Yet we have seen that tolerance of physical contact can be very different between cats and within the same cat in different environments and circumstances.

In this survey cat owners did not seem to value the 'seeking of the owner when worried', which is something dog owners feel bonds them to their pets. Cat owners did not rate the value of their relationship on this – perhaps very sensibly as it could lead to disappointment when living with a creature which has evolved to be self-sufficient and without a history of leaning on its own kind for support.

What did they find? Over a quarter of the respondents were classed as having an '*open relationship*', which meant that they had quite a neutral relationship and balanced emotional feelings about their cats. Their cats typically had access to the outdoors, got on well with other people, seemed to like their owners but did not 'need' to be near them, and seemed independent (or as some might put it, 'aloof'). These owners were less likely to own large numbers of cats but cared well for their cat and perhaps recognised the importance of choice and of keeping cats active and interested. Owners did not see themselves as a refuge for the cat nor did they rely on it to seek them out if they felt stressed.

Just slightly below this in numbers was a group which

had less of an emotional investment in the cat, perhaps not considering it as friendly or part of the family, even if the cat was friendly towards the owner but not to other people. This was classified as a 'remote association'. A *'casual relationship'* existed where friendly cats were happy with other people, didn't seem to need their owner more than other people, and often occurred in in busy households where cats often had outdoor access and maybe even visited neighbours. Both of these were characterised by an emotionally remote owner but differed in the cat's acceptance of others.

The 'co-dependent' and 'friendship' relationships had owners who were very emotionally involved with their cats but differed in the cats' acceptance of others and their need to be near their owners. Almost 45 per cent of those who answered the questionnaire were owners who felt heavily involved with their cats emotionally. This also coincided with results of owners saying the cat was not 'aloof' and also with the cat licking the owner, i.e. friendly cats which seemed to like to be near their owners. It also included cats that were happy with other people and cats, but which seemed to prefer their owner to other people, and this was called *'co-dependent'* (each dependent on the other) and seemed to happen in homes where the cat did not have outdoor access. The cat and owner played together, and the owner often stayed with the cat while the cat was eating. Authors of the paper suggest that this relationship arises from owner interaction and question whether these cats might be more likely to suffer anxiety if separated from owners, or suffered from frustration because owners controlled interactions, when the cat was fed and other things that took place in their lives. The researchers say that *'friendship'* is the type of relationship that occurs more often in houses with more than one cat, where owner and cat have an apparently friendly and warm relationship but can also function independently. This cat likes to be near the owner but does not feel a need to maintain physical proximity to them.

So what does it all mean? It seems that in this survey, the owner's level of emotional investment in the cat and the cat's sociability can lead to different types of relationship. Sociability could be seen in terms of acceptance of people other than the owner but was different to close contact and friendly interactions. The authors also noted that while many cats may seem to be 'aloof', it is not as common as often portrayed. 'Aloof' is not particularly nice word and makes us think of cats which are not friendly or forthcoming, cool and distant, and uninvolved with owners. It may be used by people who are not interacting with their cats or those who don't own one at all, rather than those who are more involved.

Early research asking owners what they might like to find in a cat which they would consider more 'ideal' than their own showed a few interesting results. Owners who had more than one cat wished that their cats were less fussy about food than people who only owned one cat. Women who owned cats that were allowed to go outside wanted their cats to be more wary of and less friendly with strange people compared to people whose cat stayed indoors – understandable really as they were presumably worried for their cats' safety. But those with cats that went outdoors wanted them to be more independent than people whose cats did not go outdoors, who also wanted their cats to be close to them. Older people were happy to wait for their cats to take the initiative to interact and were also more tolerant of any behaviour which caused damage; they were more accepting of the cat's individuality and more satisfied with their cats than younger people.

We should not take the role of cats in our lives for granted. There is nothing wrong with wanting to share our lives with cats, but examining our motivations can lead to better understanding and more thought about living together for mutual enjoyment. Let's look at some of the things we would like from our cats.

To enjoy us stroking and cuddling them

I love this diagram on the next page, which has been produced by the experts at International Cat Care and looks at the different ways in which our pet cats react to physical contact. This is not the same range of cats we saw in Chapter 2, which looked at lifestyles of cats ranging from those that don't want to be with people through to pet cats. Here we are looking at different responses from our pet cats in reaction to physical contact such as stroking.

PET CAT RESPONSE TO PHYSICAL CONTACT

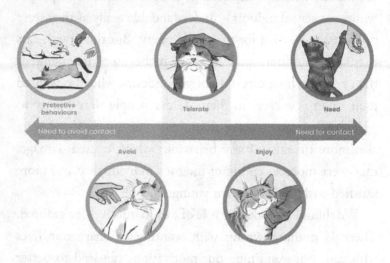

Picture courtesy of International Cat Care showing the different responses of pet cats to physical contact.

Anyone who has had several cats over the course of their lives or has friends who have cats knows that they vary so much in how much or what type of handling they want. This will depend on the many things we have discussed in previous chapters. The reactions of cats vary from 'enjoy' to 'protective behaviours'. 'Protective behaviours' may previously have been called 'aggression', but behaviour experts are, rightly, loath to tag behaviours as 'aggression'. When you look up the meaning of aggression and how we see it in our minds, it is 'the action of attacking without provocation'. Cats very, very rarely attack without provocation. And provocation itself does not mean overt poking or prodding a cat into a reaction, but may simply be a fearful cat trying to say, 'Please move away from me' or 'I don't like this situation' because someone is close by. Therefore, the phrase 'protective behaviour' has been coined to try and show that the cat is not acting without cause but may not be able to cope with the situation, one that may not feel extreme to people whose motivation is to love the cats, show that love by stroking, and persisting until the cat accepts it and feels better. Unfortunately love does not always overcome all obstacles, and fears can be deep-set. The cat reacts by trying to push people away and regain control of the situation because it does not understand that human motivation. It's the most extreme of the responses, followed by 'avoid', and probably occurs when cats can't 'avoid'. Even friendly cats sometimes avoid, just because they don't want to be stroked or cuddled at that particular time.

'Tolerate' is my favourite illustration and I think quite often our cats do just about tolerate what we are doing – they don't push us away or encourage us but put up with the contact and wander off when they get the chance! It's not a slight on us and doesn't mean they don't like us; they may just be on the way to do something else or other things are preoccupying them at the time.

Some cats love interaction and 'need' it, actively encouraging us to interact and stroke them, enjoying the physical contact. Remember too that the same cat may use different responses at different times, depending on whether the circumstances are right for it to accept contact. Cat owners would obviously prefer 'enjoy' or 'tolerate' responses to their advances, and some like it if their cat exhibits 'need' as well, but some people find this annoying.

The diagram shows reactions to stroking, but most people want a lot more than just being able to stroke their cats. People love to give hugs and kisses as signs of affection, and it is so tempting to try and cuddle our cats when we feel the need. It is no coincidence that we buy 'cuddly toys' and that they are universally attractive to all ages. Some cats may be very happy to receive a kiss on the top of the head or even to be squeezed tight, while others will not like to be invaded or restricted in such a way. Many cats don't want to be hugged and squeezed like a cuddly toy, because it's confining and limits their ability to look after themselves and to have an easy escape route should they need it. It is an instinctive response, and we may be offended that they don't realise that we wouldn't hurt them, but remember for cats it's about having choice and control. Even 'needy' cats may not want to be held tight, even though they crave some type of attention or acknowledgement – this can be rather frustrating for owners who want to sweep them up into their arms because that seems to be the ultimate way of showing that they care and will protect their cats.

Some pedigree breeds seem to enjoy closer contact and cuddling more than moggies, although of course there is a huge range of reactions to these activities across both pedigree and non-pedigree cats. Reactions may also depend on who is carrying out the action and how the cat feels at the time. If you persist, then cats may get used to it if it is done slowly and carefully, but they may just recognise your intention and react more quickly to avoid it, which is of course pushing them away rather than encouraging them closer.

Of course, cats get to know ways to act when they do want to interact with us, and they carry out the behaviours inherent to them, such as purring or rubbing against us or kneading with their paws (even if it can be a bit painful if our clothes are a bit thin and the claws reach through as they knead), as well as using meows which have been developed beyond the cat-to-cat interaction just for us. Purring and kneading are behaviours that our cats did as kittens with their mothers – the purring both encouraging interaction and signalling that all is well, and the kneading a repetition of the kitten's activity when sucking to encourage milk flow to the teat. It is of course utterly charming, and we can encourage it even more by wearing very soft, warm materials that cats love and which sometimes seem to send them into an ecstasy of kneading and purring. Presumably they are in the same mindset as they were as kittens, sucking and snuggling into a soft, warm mum who is providing dinner.

To provide companionship

Anyone who has lost a dog or cat knows how horrible it is to come home to an empty house without that creature being there and coming to greet them. Even a small cat can fill our house with its character and make it feel like a home. Companionship can be just about sharing space and having a creature there with us, or it can be a much closer relationship. Research has also shown that when we interact with our cats or dogs we experience increased secretions of the hormones found when people bond with their babies. This is not to say our cats are necessarily child or people substitutes, but they do provide us with something to care for. While we may not see them as replacements for people in our lives and social interactions, they can help to improve our mood.

To satisfy our need to be needed

The need to be needed is fundamental to humans. We want to be significant to someone or something, to play an important role and contribute to something beyond ourselves, whether it is family or community or a cause we feel strongly about. Feeling significant or having meaning is a basic human desire and a critical factor for mental, emotional and physical well-being. Pets are accepting of us, they are honest (as long as we can decipher that honestly and accurately) and can help a person to feel a sense of self-worth and love. It can help to induce contact between people, perhaps even increasing our capacity for kindness and generosity. If this sense of acceptance goes away, we may feel that we have lost a sense of purpose and direction.

We humans are social creatures and a sense of belonging

and purpose makes us feel good. Our cats may give us this chance to be needed and it may also help to explain why some people choose cats to adopt which have had very difficult lives or have illness or disabilities that require a high level of maintenance. It probably also explains why the three-legged cat is adopted from homing centres more quickly than those with a complete set of limbs! There is nothing wrong with this because there is motivation to ensure that the increased effort required to help some animals is maintained.

I am not a human psychologist, but I understand that sometimes this need can also go too far. If a person cannot do without the object of the need, a cat may feel uncomfortable to be the total focus of someone's life.

To give us attention

Cats usually provide very personal companionship – our dogs may get us out and about for walks and interactions with people, but people may not even know we have a cat. However, things have changed in the era of social media and we can share our cats and their antics online. For most people it's simply a nice way to share pictures of an animal they love and to find others who feel the same way. Social media can be used to discuss good cat welfare and to raise awareness of ways to help cats or to find them homes.

However, it can also give a platform to behaviours that may not present good cat welfare. There are many very fearful cats on the internet which still seem to raise laughs. We have seen the rise of huge followings for cats such as Grumpy Cat, which are given human characteristics and their abnormal looks are a source of amusement for many. It could be a reason for breeding more unusual-looking cats with which to associate ourselves.

Competitions such as the World's Ugliest Dog Contest bring together pictures of very unusual dogs. The most generous interpretation of the motive for the competition is that we can love animals no matter what they look like and they don't have to be beautiful. Another way of looking at it is that it's like a freak show to make people laugh and, looking at many of the contestants, it is clear that they are the result of bad breeding which resulted in distortions of body and face, strange coats and other problems. While individually we want all animals to be cared for once they are in the world, it seems to be celebrating things that could and should be prevented from happening in the first place and giving people an excuse to exhibit these dogs. The winner of one such competition still upsets me – it was a bulldog that could hardly walk because its legs were so bowed, and the anatomy of its enormous head gave it a huge gape of a mouth (no space for a nose, of course, on such a flat face) within which its tongue could not be held and so it hung out the whole time. Part of the prize was a trip to New York to show off the dog some more – a dog that could hardly walk or breathe without difficulty, something that would be exacerbated by stress or heat, and it had to travel to a far-off city by plane or car. Where is animal welfare in this situation?

A similar ugly pet competition in Asia chose a Chinese Crested dog and a Sphynx (a hairless) cat as ugliest. Other cats included a Scottish Fold and a Persian cross – all cats bred deliberately by people. I almost cried when I looked at a YouTube video of Wilfred cat, a social media 'treasure' – it has been bred with many inherited problems so that his eyes bulge from a face so distorted by breeding for extremes of conformation that he must find it hard to eat. The video starts by saying, 'There is nothing wrong with Wilfred – he is just ugly'. There is a cause for problems such as this – it's people-shaped, and we should be aware that these animals' issues are preventable and not to be celebrated. We need to care for animals that result from irresponsible breeding but also try to prevent it happening in the future. In Chapter 7 we look at some breeding practices that are potentially abuse of cats.

Dressing up cats is another way of gaining attention for people and is something cats probably don't 'enjoy' and at best probably 'tolerate'. Some will be actively distressed and many freeze or fall over if they are dressed up – where is our respect for this lovely species? Others, such as Scottish Folds, which suffer from arthritis, may not want to move or react because of the pain or discomfort caused by the condition. Cats are not dolls to be dressed up according to the season or the celebration.

To think and feel in the same way as people

There is a belief that giving our pets tags such as 'fur babies' or ourselves the title of 'pet parents' will help us to be empathetic to our animals and treat them as members of our family, and thus with sympathy and responsibility. I have a particular dislike of both of these names because there is a danger that they permit us to think that our cats are human, with the same thoughts and feelings as us, and that we should treat them in the same way as our children. It could be seen as an easy way to interpret our sometimes mysterious and complex cats, but it can also be an easy way to misunderstand them because they are not people and have different wants and needs, as will be shown in Chapter 5. It allows us to have very emotional responses to animals, thinking they are lonely or jealous or need to be bought expensive items (a position used by makers of pet products encouraging people to prove their love by spending lots of money on their pet) or that love in human form can overcome any problems.

Anthropomorphism, which is the attribution of human feelings to animals, can improve empathy (the ability to understand and share the feelings of another), but if we are misinterpreting these feelings then it may also detract from prioritising the needs of cats. It's about respect for the species and the individual and we may have to adapt our behaviour if we really want to improve our cats' lives – sometimes we don't actually want to know, because it doesn't suit us and we may have to change our attitudes and actions.

To react positively to our 'love'

We may not want to think like this but, as also discussed earlier in this chapter, research has shown that our affection for our cats is linked to our perception of their affection for us, how predictable our cat is, how curious it is and how playful and clean, so we have quite high expectations and do seem to want to have some sort of payback for our love. People who are highly attached to their cats are more likely to follow them around and want feedback from the cat. We want our cats to like or even love us – but at least we don't expect them to be obedient or to protect us as we do of their canine cousins!

A survey of owners found that indoor cats were rated highly if they were more active, interacted with their owners and initiated interactions. Results also reported that outdoor cats seem to rub more against their owners more than indoor cats, especially when they come inside, perhaps realigning their connection by greeting and exchanging scents. Outdoor cats were rated as less curious than indoor cats. However, this survey raises lots of questions – not just about the behaviour of the cats but about the interpretation of the owners. Cats that went outside may have been just as curious as indoor cats, but satisfied their curiosity about all sorts of things and used up their energy before coming in, greeting their owners and then resting. Indoor cats may have to focus their inquisitiveness on owner activities because things don't happen spontaneously and out of their control, as they do outside. This would probably focus the cat's attention more on its owners.

The saying 'love conquers all' may be romantic and optimistic, but it's not necessarily true of our cats! If you've understood the previous chapters about the varying levels of interaction that different cats prefer, you might have worked out that even copious amounts of love may not get a cat to change its personality or relieve it of its anxiety towards people or life in general. We may want to believe that if we just keep loving these cats they will acquiesce and become pets that give back in return by enjoying handling. There is no doubt that love can drive us to give, but it is the nature of love which is important. Good love incorporates respect and patience, and the proof may be in our ability to love a cat from a distance without physical contact and not pursuing it for this type of interaction. If the cat makes a move towards us, then we can react, but pushing the cat too far or making it feel anxious is not love. It may take years for an anxious cat to make even a small move or just not move away, but even that is to be appreciated.

We can also love our cats too much by over-feeding, partly because the act of feeding can be an interactive and rewarding time with them. Even a street cat can be interactive to some extent, making overtures if it feels that this will encourage people to put food down for it — it may not want a stroke or a cuddle, but it is a period when there is communication between cat and human.

To be happy to be with us and our friends and family

Some people quite like it if the cat likes only them and nobody else because it makes them feel special. Some cats, probably those who have had less than optimal interactions with people in their early months, do seem to be happier with one person and don't spread their love to anyone else. Cats are a much more private pleasure than dogs in that our friends or acquaintances may not meet them unless they come to our homes. Most people, however, like to show their cats off to friends and family and love it if they are happy to interact and come for a stroke, or even target our friends to sit on their laps (as long as it's the guests who want to interact!). People often mention that cats choose the person who doesn't want to interact with them to approach. Whether this is because people who are not fond of cats may not be looking directly at the cat or actively trying to get its attention and may be sitting still and quiet, which is all interpreted by the cat as unthreatening and so quite appealing, we are not really sure. It is rather annoying though for those who don't want cats to approach them!

To want to be with other cats

As we have seen, the cat's ancestor is the African wildcat, a solitary species seldom meeting its own kind except for mating or raising kittens. Our own domestic cats are not quite so definite in their desire not to be with other cats – they are more flexible, at the same time not having the absolute need for companionship that we see in dogs and people. We also don't know, even with a very sociable cat, whether or not our absence is actually felt as loneliness or they accept being on their own quite happily, and then adapt to the good things which come when owners are around. We know cats vary their activities and the time they are awake to fit in with their owners, especially if they are indoor cats that see little variation in the day in a quiet house, compared to the cat which can go outside and find adventure. However, if we think our indoor cat may be lonely and want to get it a 'friend', it can be very difficult for this only cat in a very fixed environment to have its limited world turned upside down by introducing another cat that it may not like. If you are planning to get a cat to keep indoors it may be better to go for two kittens together to start with, and give them space to get away from each other if they need to as they grow up. However, this does not guarantee they will remain friends and we need to keep an eye on their relationship.

Some cats – usually breeds such as Siamese or Burmese, which can become very attached to their owners – may suffer when left alone. Know your cat – don't assume it wants another feline around just because you do, or because you leave it alone sometimes. Some cats are perfectly happy like this and would be very upset at the introduction of a 'friend'. Also, don't assume that just because a cat has got on with one other cat which has unfortunately died that it will welcome another cat – think how you might feel if someone decided on a companion for you, with whom you might not get on or can't get away from.

Remember the territorial nature of cats as discussed in Chapter 1 and how strong that instinct can be. Cats are individuals with their own likes and dislikes, and having another cat forced on them can be very stressful for either cat. It also takes time and patience to integrate another cat as it needs to be done very carefully. Some cats do like being with other cats; kittens usually get on well together, but as they reach social maturity at around two to four years old, they perhaps need more individual space, especially if they are kept indoors. Some houses are not suited to several cats kept indoors – those with small rooms or those with lots of narrow staircases where cats in conflict can prevent another passing may not help with relationships.

To be clean and not destructive in our homes

One of the reasons a cat is such a popular pet is probably its cleanliness. Cats may use the great outdoors or a litter tray indoors and usually there are no problems. Cats are generally very clean, grooming themselves and seldom bringing in mud unless attached to a long coat. They don't smell, even as they get older (unlike their canine cousins!). Cats are also unlikely to chew things because they are suffering separation anxiety when their owners are not there – something dog owners need to address and find solutions to.

The 'destruction' aspect with cats is mostly linked to scratching things in our homes. Cats have an innate need to sharpen their claws, combined with using secretions from their paws as scent messages and visual marks. Many cats do scratch indoors, and most owners tolerate a certain amount – there are things we can do to direct scratching onto scratch posts, but it may not be 100 per cent successful. However, the answer is definitely not declawing, which is fortunately banned in many countries, including the UK, but not in others – see Chapter 7 for more discussion of this horrible practice.

When you read this, and the next chapter about what cats want and need, you will see that people's needs and wants sometimes don't quite align with those of cats. What we need to be able to do is to understand where things don't quite align and work out what we can do.

5.

What do cats need and want, and how does this align with what we want for them?

WE THINK OF our cats as members of our family, so if someone asks what we want for them, perhaps we might answer in the same way as we would, say, for our children or relatives – to be healthy, happy and to have choice in their lives. How does this match with what cats want or need? As we have seen, what may be one cat's pleasure may be another cat's horror and what we want for our cats may not be what they might want for themselves. We are, of course, somewhat guessing what cats do need and want using our limited understanding of their behaviour.

However, it is worth trying because large numbers of cats are given up to rehoming organisations every year and 'behaviour problems' are one of the main reasons cited by relinquishing owners. Reasons seem to suggest that owners' understanding of the needs and behaviour of their cats is poor, and vets also feel it is often because owners do not understand their cats' needs and are stressed. So a better understanding of a cat's need should make expectations realistic and perhaps allow owners to help their cats to feel comfortable with them.

Health is something we can work towards – our knowledge of cat disease and treatment and prevention of disease is developing well, thanks to the early work of International Cat Care and other early pioneers for cats. Because cats have become such popular pets, and because they have always had people who have championed them, we can now provide good healthcare, treatment and nutrition. There are always new things to learn but we have come a long way in the past sixty years in being able to ensure our cats can have good treatment and pain relief if they are unwell.

We would like to think our cats are happy, but what does 'happy' mean if you are a cat? In this chapter we will look at what they need and what they might want – and presumably if we can fulfil those needs then they will (hopefully) be happy. We know ourselves that to be happy we have to have a structure in place that makes us feel safe and comfortable so we can relax and enjoy experiences and relationships – for cats it is no different. We might think about cats playing or purring as being a show of happiness, but they won't do those things if they don't feel secure, are not hungry or thirsty, in pain or feeling unwell, are comfortable and are not anxious. Happiness is the pinnacle of the layers of needs which build on all of these things. A cat won't want to consider giving affection or being friendly if these other things are not in place. It's pretty difficult to define what happiness is in people, but it is something we would like to attain for ourselves and our cats.

Perhaps it is easier to explain these factors as cats feeling safe in our homes, eating well, being healthy and feeling fit, and being comfortable with the type of interaction we have with them and the environment they live in. We don't want them to feel 'stressed' or frightened. It does need us to 'think cat' rather than attribute human feeling and needs to our cats, and to be accepting and understanding of their particular personalities.

The cat's point of view

We know that there is a difference between the way cats live without people and the way we want them to live with us. Most people may think that having a demonstratively loving, cuddly interactive pet cat that is obviously relaxed and happy to live alongside them is 'normal' and that all cats are like this. Many cats and people do indeed have a brilliant and mutually enjoyable relationship – if you have one of these, then enjoy! Is this because owners are understanding of their cats or because these cats are relaxed with human needs, or a bit of both?

Cats living with us get the benefits of shelter, security, a continuous source of food, warmth and comfort. They also get our 'love', but that is a very broad word, as we've seen, and, just as in human relationships, can vary from a respectful and mutual trust with both sides giving each other space to do their own thing and enjoying physical interaction when both mutually want it to a controlling, or disrespectful relationship which does not take into account the reactions or needs of the other.

At one end of this spectrum is living with a cat with very little interaction, and, to be honest, many cats can deal with this very well, except perhaps cats which really want our interaction. For many cats that have not had the genetic or early experiences to make them into very sociable cats that really enjoy being tightly involved with people, this situation may indeed be ideal! At the other end is a human need for control of behaviour and of all the risks that may come in a cat's direction, as well as a need for physical interaction even when it's not welcome or to a level which the cat is not comfortable with. Do we interpret the cat's behaviour in human terms because that is what suits us, rather than understanding it from the cat's point of view, because we don't want to change what we do?

If we could look at things from a cat's point of view that would be a good start! What drives and motivations do cats have that are not suppressible? How have their earlier lives affected their adaptability to allow them to fit in with people?

And what individual preferences do they have? Of course, cats cannot tell us directly what they want or need, so we have to interpret their behaviour in the light of what we understand about cats in general and about our own particular cats.

All cats have common needs of food, water, a place to feel safe and rest, and a toileting area, but even this may not be straightforward as the cat's needs may also be influenced by its confidence and other people or animals in the environment in which it lives. Cats need an outlet for behaviours that are part of their natural repertoire, such as scratching and rubbing, and an outlet for hunting instincts (which may be through play). They obviously also need freedom from discomfort in the environment and from the discomfort of ill health, pain or injury, and from fear and distress. Care needs to take into account the cat's mental wellbeing as well as its physical health. Some animals, such as dogs, need company; others, like cats, need to have the opportunity to get away from other animals (often from other cats) if they wish.

Those are the basics, and some are very obvious, but even the simple things may be more complex than we think. We broadly know the basic things cats need to survive and thrive, but cats may enjoy having a choice of more than the basics – selecting the best seat by the fire or a particular favourite food. And while there is science available for understanding the 'needs', we do have to try and interpret the 'wants' of our own cats.

Self-reliance, choice and control

As we saw in Chapter 1, the cat is a self-reliant animal (whether we help it or not), influenced by genes and experiences that may make it happier to accept our help and care. Its evolution makes it vigilant about its safety and open to opportunity to hunt and thus eat. In order to be self-reliant, it needs to have control over its environment and be able to have choice about what to do. If the cat perceives danger, whether it is real or not, it will need to be able to act to get away from the threat. Some of this may come across as illogical to us because we are not 'thinking cat'. A behaviour which usually amuses us is when we close a door and the cat tries to open it, and then when we open it the cat does not go through – for us that doesn't make sense and it seems as if the cat is just being difficult for the sake of it. For the cat it is about an instinct to have an exit, just in case it needs it – it just wants to have the option! It is not stress, just a bit of frustration if it is prevented from acting in a way to help it cope with the feelings it has!

More serious lack of control and therefore more stress can develop if the cat cannot get away from another cat it doesn't get on with on a long-term basis. The cat is living with anxiety and fear and there is nowhere to hide to relieve this. Another example may be a cat which has been used to going outside but for some reason (which may be medical or the owner's choice) is now confined indoors – it may become frustrated, and this can result in a change of behaviour people may feel is bad behaviour, such as soiling indoors or hiding or reacting in what seems to be an aggressive way.

To feel secure

A cat values its territory highly – it is an area to defend for food resources but also one where it has somewhere to rest and shelter. Just because cats live in our homes and we feel they are secure, it doesn't mean their instincts for safety and refuge are removed. Even in our homes cats are looking for a secure territory with plentiful food, a safe space and a toileting area. We may think the inside of our home is indeed secure, but is it secure in cat terms?

For a cat that goes outdoors the house is part of a bigger territory but is the very safe bit at its core, and while it may have to deal with things that happen in the outdoor territory, indoors is probably regarded as much less challenging. However, when we bring in a new cat as a 'friend', or when a cat 'visits' from outside and comes in through the cat flap, that security is threatened. For the totally indoor cat its territory is smaller and more focused and, if a new cat arrives, it takes over the entire territory and is a huge intrusion. I hear many people saying how lovely it is that next door's cat visits their cat and eats its food. While you can never say that all cats will find this stressful, most cats would not enjoy this even if it does not end in a face-to-face fight – they may just hide away for a while until the other cat has gone. If it comes in and the cats curl up together to sleep, then it would seem that it is welcome, but I suspect this is quite rare! If someone walked in off the street into our home and ate our food, even threatening us in the process, we would be upset and anxious about it happening again. If the door is not locked and we know they could come back at any time, we would worry, and that worry is almost worse than the reality, which may not be physically damaging but shatters feelings of safety and security, allowing no relaxation.

What do our cats need us to do to improve their security? We can think about having a selective cat flap which works for our own cat's specific microchip to stop others coming in. We can ensure we carry out proper introductions of new cats. We can also recognise the importance of familiarity of smell within our cats' own homes.

The scent-orientated cat is as familiar with the scent layout of its home just as we are familiar with how our homes look. We know if someone has changed things around just as cats know if another cat has been in. Being aware that a change of furniture/carpet will change the look and smell profile of their safe house may upset cats' feelings of security is important. And while we have to change our homes to carry out renovations or add new carpets or furniture, it is important to know how our own particular cats will react. Confident cats will adjust easily and we may not notice anything as they have taken the alterations in their stride; more nervous cats may take longer to relax with the new sensations in their home. The totally indoor cat is much more likely to be affected by change because it has more constancy in its life than the cat that goes in and out, which may be dealing with challenges and changes outside and constantly having to make decisions as to what to do. Small alterations may seem massive to the indoor-only cat: these include people, other cats, babies, routines and furniture.

The need to find secure places in which to rest will vary considerably with the level of activity and number of people and other animals which may inhabit the same area. If there are young children, dogs or other cats that may make life difficult, a cat is likely to try and go upwards to find a place of sanctuary. And although other cats may be able to follow, it does give a space where at least the cat can notice a threat coming and from which it can defend its spot. If you own a nervous cat, you may notice it takes to high areas a lot; tops of wardrobes are a favourite. It may also hide under things so it can watch from a safe place.

For those inbetweener cats that people care for (see Chapter 2), owners can have a more distant relationship and provide shelter outside in sheds or provide boxes/shelters that are warm and secure, giving the cat space while providing comfort, food and shelter – more on this in Chapter 8.

To feel safe

Any cat lover would say yes to the question: 'Do you want your cat to be safe?' However, with animals such as cats which have traditionally come and gone from our homes as they please, complete safety can be a difficult issue. Should we be removing all risk and putting safety at the top of the list for cat welfare? Many people might say yes, others will not agree. Owners who rate their cats' independence highly have to balance the risk of going outdoors with the value they place on their cats' right to free choice of lifestyle.

For some people, even the tiniest risk to their cat's safety is something they cannot accept, and the cat must be kept inside all the time. This will remove some, but not all, of the risks of injury. While an indoor cat cannot end up in a fight with a cat that has an infectious disease, being confined can add risks of a different type. And it may be that cats kept totally indoors may investigate their environment more closely and get themselves into trouble with things they might not be interested. An example of this is cats chewing indoor plants, because it is the only greenery available, and they have a need for some vegetation or because they are curious, something they would not touch if they had access to the outdoors. Outside they would chew some grass or other plants perhaps as medication or for other reasons we are not really aware of.

There are different risks: those in our homes, which we can control somewhat, and risks outdoors that we have less control over. We know cats are curious, especially when they are young and inexperienced, so we do need to think about the risks inside our homes. These include crawling into small spaces or on to high levels. And if the space is warm, such as a tumble dryer or washing machine, then it is even more tempting. Kittens' curiosity will have them climbing up on to hobs which might be warm or into ground-level fridges when the door is open; getting into cupboards where bleaches and cleaners are stored; playing with needles and threads used in sewing or even chewing wires. Indoor cats and kittens may sample potted plants, or bunches of flowers that may contain plants such as lilies, which are poisonous to cats. Cats kept in high-rise flats may access balconies or open windows and fall if distracted by birds.

Security and safety sometimes work against 'choice'

So owners will need to be vigilant and see dangers inside too. But there may be additional and different risks to cats that are kept indoors, and these are often associated with mental wellbeing. They may be seen as less important than physical risks but should be seen in the context of the cat's wellbeing and not just safety. Outdoor risks may come from cats, dogs, people or accidents from being in the wrong place at the wrong time, poisoning, disease or parasites. Indoor risks may include issues caused by cats being bored, frustrated or fearful of change; cats may put on weight through lack of exercise or becoming too reliant on their owners; risks from indoor hazards; and, of course, escape but not being street-wise when they get out. This also emphasises the need for them to have identification by microchip in case they accidentally get out and have no street wisdom.

For some cats, being outside in the area around their homes brings high risk of injury on roads or from other animals, or they live in high-rise flats, where allowing cats to go out is an impossibility. Whatever the risks, cats may not appreciate the limitations put upon them. It would be unfair to take a cat that has had outside access during its life and keep it in, and so better to start with a kitten which is kept inside, rather than reaching a point when vaccinated and neutered it can go outdoors. But depending on the cat, it may still be frustrated with this setup.

People now build fences or enclosures around their gardens or part of the garden, which gives the cat safe access to some of those things that allow it to express its natural behaviours. This is a great compromise.

Food and drink

Our pet cats don't usually go hungry or thirsty, but they can be quite opinionated about what they like! We know cats have special requirements nutritionally and can't be vegetarian or vegan. They have maintained their hunting abilities so, if they go outside, they can supplement their owner-fed diet if they need or want to. Their senses are developed to be discerning about their meat diet (for example, they are not thought to be able to discern what we know of as 'sweet', as their natural diet would not include sweet items). Perhaps combinations of amino acids, the building blocks of proteins, give them different sensations when they eat different prey. Given the choice, cats will choose food that has a high meat and fat content, a strong smell, a mixture of soft and crispy textures, and they prefer to eat it at body temperature. The perception of taste and smell will affect whether a cat finds a particular food palatable and will affect how much it will want to eat. So they don't particularly like cold food from the fridge and may like variety unless they are feeling stressed, when they may revert to things they recognise and know (a bit like us when we revert to tomato soup and toast when we don't feel well!).

And, like us, they may be attracted to eat or drink things that are not particularly good for them or may not agree with them. Traditionally cats are given milk, but beyond kittenhood may not produce enough of the right enzymes to digest milk anymore, so drinking it upsets the stomach and causes diarrhoea. We also have to remember that cats are good at asking us for food and that if we provide too much or the wrong type it may cause obesity or illness. So, we don't always have to give in to their wants! However, what we do need to do is be aware of anything in their lives that is driving them to eat more or less than they should. We have developed very palatable foods for cats that make it easy to intake too many calories, but they may also eat because they are bored or stressed. Do we as owners use food as a currency for interaction, loving the attention we get when we feed cats, and give them too much? If they are losing weight, are they ill or is there something putting them off food?

Many people have more than one cat, each with its own preferences: some may be fussy and only like one particular flavour or brand of food; others will eat anything, anytime and even go for foods that are not in the normal cat repertoire, such as vegetables or crisps. What a cat likes or dislikes is determined by what it has experienced in its life and the associations it has with those foods. Some cats become fixed on one particular food which may not be the best for fulfilling their nutritional needs and it can be very difficult to change them. The theory is that if you want to change a diet, you very gradually mix more and more of the new diet into the old one until you have moved over entirely to the new food. Getting cats to be happy to eat both wet and dry food may help if there are problems about diet in the future. Remember, too, that it matters where food is placed, as outlined in Chapter 2.

To be healthy and avoid pain and injury

Good health is a blessing for cats, as for people, and it involves a combination of prevention of disease, early detection and treatment of problems, minimising risk of injury in the things we do – and good luck with our genes! So, what can we do? Thinking long term about what our cats will need and planning ahead will help us to keep them in the best possible health. By undertaking routine preventive healthcare through all stages – from kittens through to old age (see Chapter 2) – we can ensure our cat stays as healthy as possible. We can also prevent problems arising or maximise the chance of early detection of any illness.

A good start to health and wellbeing: A good start in life can't be underestimated. Choosing a healthy well-socialised kitten is a great foundation for a healthy life. When we say 'healthy', we mean physical health and mental health and as we have seen in Chapter 2, those early weeks are very important in preparing cats to be pets and live happily with people. So, finding out how a kitten has been brought up before you take it on is very important. A very useful guide has been put together by many of the UK's animal welfare charities, which got together as The Cat Group and developed what is called The Kitten Checklist. It is available online and looks at what to find out before you go to see a kitten and what to ask and observe when you are there. It covers things such as what to look for in physical health and how to ascertain whether the kitten has had a good human experience in those vital early weeks, which will make it relaxed and happy with you as a new owner.

Vaccinate: International Cat Care began in 1958 because there were no treatments for cats when they were ill. People knew that there were diseases that were spread between cats, but there were no vaccinations. Just as in the early days of medicine before vaccinations were developed and many (especially young children and older people) died from a wide variety of diseases, infectious diseases took a great toll on kittens and cats. These days we have almost become complacent about these diseases because they are well under control. COVID has now reminded us that we are vulnerable if not protected and we should never take that protection for granted. Therefore, vaccinating kittens against cat flu and infectious enteritis is vital, even if they do not go outside, because we can bring these diseases into our homes on hands, shoes or clothes, and they can be severe if the cat is not protected.

Vaccines which protect against these diseases are known in the veterinary world as core vaccines. Other non-core vaccines are only given to cats if there is a genuine risk of exposure to the infection and if vaccination would provide good protection. Decisions regarding the requirement for non-core vaccines will be based on the cat's age, lifestyle and contact with other cats. You should always discuss with your vet what vaccines your own cat may require. In some countries vaccination against rabies is mandatory and there are very good reasons for this in terms of people and cat safety. Additional boosters throughout the cat's life will ensure these diseases will not be killers.

Treat for fleas and worms: Regular treatment against worms and fleas will help to keep your cat in tip-top condition – these days there are more and more products available which are easy to give to cats, either palatable or mixable with food or in other forms (such as products which can be put on the skin at the back of the neck – called spot-ons), that make keeping our cats free from parasites easier. Cats which go outside and hunt will particularly need to be treated as they are more likely to pick up parasites from the prey they eat.

Provide a good diet: We know cats can't be vegetarian or vegan, so to keep them healthy we need a diet with meat components. Luckily many petfood manufacturers provide diets with which we can keep our cats well. Trying to feed a balanced home-made diet to cats is very difficult and can often result in deficiencies, so it is probably easier to rely on the research that has been done by pet food manufacturers to provide a balanced diet.

Neuter: Cats are very successful at breeding, but it brings with it certain behaviours and additional risks to health. Neutering takes these away and makes it a great deal easier to live with our pet cats, as well as removing the responsibility of finding good homes for kittens. Owners may put cats in a situation that raises the chance of injury caused by conflict by keeping individuals together who do not get on. Or by keeping cats that are not neutered in an area where there are other cats, raising the risk of fighting, which is almost inevitable with hormone-fuelled territory disputes. Neutered males stay nearer to home and are much less prone to getting into conflict or having accidents.

Stay vigilant: We humans are used to having check-ups with our doctors in many different stages of our lives, from baby checks through to middle age check-ups and various tests along the way. The same goes for cats. In Chapter 2 we looked at the different life stages of cats and for best care, cats need checks-ups during these, just as we do. It is our responsibility to care for our cats both in prevention of disease and illness and in recognising illness or pain and acting to help them.

Be able to recognise signs of illness: Even if we vaccinate and treat our cats for parasites regularly, they, like us, can become unwell. For cats, signs of illness can be subtle and vague! It is obvious if cats are vomiting or have diarrhoea, but mostly they go a bit quiet, seem lethargic, dull, lacking in energy or listless; they may not eat as they normally do, and they may try to hide away.

Cats may also lose weight, so again it is useful to know what your cat weighs when it is well. Knowing your cat's optimum weight (you can check this with your vet) will help you to be able to judge whether it is losing weight, or gaining. And we have to consider the weight loss in relation to the size of the cat. A cat does not weigh a great deal, so losing several hundred grams may in fact be quite a large percentage of its body weight and be significant, so we need to consider weight in cat terms.

Other visible signs may include discharge from the eyes, ears or nose, a coat which is dirty or 'staring' (instead of lying flat and being smooth and shiny, hairs are standing up, separated and dull), itchiness or parasites, pale or yellow gums, bad breath, or blood in faeces or urine. Behaviour changes may show if they are eating less, perhaps drinking more, not wanting to move or jump. Urinating more or less, seeming to be constipated or having blood in faeces or urine are signs that the owner of a cat which uses the outdoors as a toilet area may miss unless the cat urinates indoors. This can happen if a cat is suffering from an equivalent of human cystitis when it may well try to urinate small volumes indoors in lots of different places. Anyone who has suffered with cystitis will understand the constant need to try and urinate because the urinary tract and bladder are inflamed.

An unwell cat may also have a hunched posture and be quiet. Often we may feel something is not quite right with our cats and we should have faith in these feelings and try to figure out what is happening. This is where knowing our cat's normal behaviours is important – so we can pick up signs which may be quite subtle but may allow us to catch a problem earlier than if we wait for a big, obvious sign.

Be able to recognise signs of pain: Cats feel pain in the same way as we do, but they may not act in the same way as us and are sometimes their own worst enemies on this front! As animals that do not cooperate to hunt or help each other if they are weakened, and as small animals which may be prey themselves, showing signs of pain will not bring support from others but may show them to be weak and more susceptible to becoming prey themselves. Researchers have recently been studying cats' faces and body language to be able to bring to light what are subtle changes, but once 'seen' can make people more alert to their cat's health and feelings. It won't surprise you to know that the signs of pain are similar to those of illness – the cat grooms less or focuses its attention on a certain area which may be where the pain is coming from, or maybe just to settle itself down with grooming (a cat in pain may also purr). It may be sensitive to touch in that place, too or push its owner away or react with meowing, hissing or growling, or even meow in a way that seems different to its usual cheerful meow.

The cat may move around less and be less active, be quiet and seem withdrawn, perhaps hiding, and definitely not want to play and maybe not to eat either, or only eat certain things. It may urinate or defecate inside the home or outside of the litter tray. The cat may walk in a strange way or be reluctant to move or jump or go upstairs. Indeed, it may just be that the cat is not doing what it normally does and this can alert owners to something being not quite right. It may also be that a normally friendly cat does not want to interact or be stroked and is avoiding people. The cat may look tense – scientists have now studied the facial expressions of cats in pain and have come up with pointers which include narrowed eyes, ears held further back, a tense muzzle and/or whiskers projecting straight out from the sides of the face, rather than loose and curved. If you google 'feline grimace scale' you can see some of these facial expressions and learn to recognise them for yourself.

For the veterinary experience to minimise stress or fear: None of us want to think of our cats as frightened or stressed. In previous chapters we have looked at what we can do to try and make them feel secure in our homes and how to produce more confident cats in the first place. Another aspect of having pets is ensuring they are kept healthy by taking them to the vet. We know many people don't like to do this because they find it difficult to get cats into baskets, they know cats don't like it in the car, and they dread going into a veterinary clinic because there are dogs in the waiting room.

International Cat Care has been pioneering in this area and developed a programme called Cat-Friendly Clinic (and Cat-Friendly Practice in the US). It is driven by understanding that there are unique difficulties in bringing a cat to a veterinary clinic and aims to provide a high standard of cat care by understanding the needs of cats. These include a waiting space in which there are no dogs, approaching and handling cats gently and with care and respect, developing hospital areas that are quiet and restful, and using equipment which is adapted for cats. Achieving Cat-Friendly Clinic accreditation involves all staff, including receptionists, nurses and vets. Each accredited clinic also has at least one 'Cat Advocate', someone who ensures the cat-friendly standards are adhered to, and who is happy to talk to cat owners about their concerns. We would be unrealistic if we did not recognise the barriers to taking cats to the vet, but irresponsible if we let them affect our cat's health by delaying or avoiding it, so finding a Cat-Friendly Clinic with staff who are passionate about cats can be a very rewarding experience for owners and a good one for cats. Instead of it being a frightening place to visit, it should be a welcoming place with people enthusiastic about cats and their physical health and mental wellbeing – the good ones will be delighted to tell you what they do.

This will also extend to how to give medication to our cats if they need it. We know that owners are not brilliant at giving tablets to cats (myself included) and vets are aware of this, too. There are now many different ways in which medicines can be presented to make them easier to administer and so ensure not only that the treatment is given, but that the relationship between us and our cats is not affected by having to fight with them. A survey by International Cat Care, which introduced the concept of 'Easy to Give' medicines, showed that many owners found that giving medicines did affect how their cats reacted to them during the treatment and were dismayed that the cats tried to hide when they wanted to be giving them comfort and security. If a medicine can be hidden in food, has a flavour that is attractive to cats or can be given on the skin, it may be that the cat does not notice and the stress is avoided.

Being with people and not 'lonely'

People are socially obligate – most of us don't do well without the company of other people and being lonely is something we see as a bad thing. For us, loneliness happens when we do not get the feeling produced by having contact and relationships with other people. People are different in their needs, too – some enjoy living alone with little contact, while others may be within a group but still feel lonely if they don't feel valued or cared for. But, as we've seen with cats, choice is important: if we choose to be alone, we may be happy, but if we crave interaction, then lack of this may make us feel lonely. Solitary confinement is seen as a punishment. For people it is a complex feeling associated with negative emotions.

However, we should not make the same assumptions about cats. We may feel guilty when we leave them alone while we go out to work (especially since lockdown and more home working) and don't want them to feel isolated or in need of company. This feeling may be proved right because our cats may be waiting for us at the gate or run to us with their tails up and purring when we arrive home. We are of course flattered and assume they have been pining for us all day.

We miss our pet cats, but do they miss us? Of course, that is a pretty hard thing to find out, but researchers have been trying! One of the positive things about cats is that they seem to be happy enough when we leave them alone when we go out – unlike many dogs, which suffer from separation anxiety. Researchers have looked at the effect on cats of being separated from their owners. They chose owned cats with a mostly indoor lifestyle (those that were not entirely indoors had supervised outdoor access) and that had access to dry food and so were not waiting to be fed when their owners returned (and their reactions were not associated with being fed when they were hungry). They found that cats spent more of their time resting when their owners were away for four hours compared to if they were only away for thirty minutes, which does not seem to suggest they were stressed. But they did purr and stretch more when their owners returned after four hours when compared to returning after only thirty minutes. As we saw in Chapter 1, purring can be a way of cats asking for interaction or to be given attention, and so purring more after a longer period away may show that cats are keen for attention, and perhaps that they are happy that their owners are home and normality has been resumed.

I wonder whether the purring was the quiet type of purr or the demanding type of purr we've mentioned? What about the stretching? We have not mentioned this before and it is not thought to be used when greeting, but perhaps it is just because they have been resting more and need to get themselves ready for action again – cats are good at stretching and getting their wonderfully agile bodies to work fluidly.

How important a person is to a cat will depend on all of these things too. But cats like to know what is happening – we say they are 'control freaks', liking to know where things are, that doors are open, and they can relax knowing what the day's routine is and are seemingly a bit unsettled if that changes. So even if the cat is quite distant and does not really want to interact physically with people, it may be that it is reassured by routine and knowing that the people who provide food and shelter are around. And, of course, the cat's normal lifestyle will make a huge difference too. The researchers in the work mentioned above chose cats that lived entirely indoors or had restricted access to outside – thus their lives would be more influenced by the presence or absence of their owners than cats with free access to outdoors, which could be so absorbed in their outdoor activities that they didn't even notice their owners were not there! Indoor cats may be ready for some action for the day and bored if left for a long time, ready to do something.

In reality, the cat has probably found a nice warm spot on a radiator or bed and happily slept the afternoon away, moving occasionally to follow the sun or the warm spot, having a snack and occasionally using the litter tray or going outside to do a bit of patrolling or hunting. If you don't leave food for the cat and it has to wait for you to be fed, then your return may well be reacted to with great enthusiasm, and you will be followed around until you come up with the goods!

Many cats really enjoy attention from people. Most cats are highly adaptable and can live very happily alongside us in as close a relationship as suits both parties. When cats want attention they will seek it out and often ask for it by rubbing themselves on us where they can, making miaowing or chirpy noises, running towards us with their tails held up straight or simply plonking themselves on our laps and purring loudly. How the interaction proceeds depends on how rewarding it is for person and cat. Some cats have a large need for attention and are very people orientated – this can be common among some of the pedigree cats such as Siamese or Burmese, which have been well socialised as kittens. Others only need attention and interaction when, where and for as long as they choose.

A cat's natural diurnal rhythm is quite flexible, with an ability to be active and highly effective at dawn and dusk – hence their common desire to get us moving early in the morning! How we react to the approaches of cats at any time will affect how they act with us in the future and how we learn from each other. That is not to say that we have to get up if they want us to interact at the crack of dawn, but that we understand the motivations and provide simple reactions or ignore some demands with consistency while welcoming others.

Being with other cats

Our knowledge of how cats evolved from wildcats to our domestic cats shows that they are not quite as solitary as their wild ancestors and can form bonds both with other cats and with us – sometimes referred to as being socially flexible. One of the 'five freedoms' we talk about with animal welfare is the need to be with or away from the same or other species. When writing about cats in general and trying to pinpoint what they want, the wide range of individual characters of cats always ensures that it is difficult to make sweeping statements. It is hard to say what cats want in terms of companionship with other cats. We know they don't need to be part of a pack or group in the same way that dogs and people do, and that many of their natural instincts are to push other cats away based on their need to protect a territory. Even if they have been friendly with one cat, it does not mean that they will take to other cats. Think of the way we keep cats and how we often have several (myself included) and expect them to share territory. The need is to understand what your cat can stand, how your circumstances allow it space, etc. – to read your cat's wants. The difficulty is that this may only become apparent when you get another cat.

For cats that have free access to our homes and our gardens, finding some space may not be difficult (although in areas where there is a high density of cats, they may visit each others' gardens and invade the space of others). Some indoor cats may become frustrated or are in conflict with other cats from which they cannot get away – and this may result in behaviours owners may term 'problem behaviour' but which cats are turning to naturally in order to try and cope with the situation. These could range from spraying or urinating outside the litter tray to being defensive (often seen as aggressive) and not relaxed, suffering from prolonged stress that may also affect health.

To be able to carry out natural behaviours

We talk about natural behaviour, but do we really understand the drive for animals to do certain things? I don't think we do – we think they are 'natural' but do not necessarily understand the motivation to carry them out and what a difficult position we put animals in if they cannot do these things – these are true needs. Just because we provide food doesn't mean that we remove the need to hunt; just because the cat doesn't hunt doesn't mean it won't feel a strong need to keep its claws sharp; just because the cat has no coat, or the owner grooms it, doesn't mean it does not automatically need to do that itself. So, what are the behaviours we could consider 'natural' for the cat? These must include grooming, toileting behaviours, claw sharpening, hunting, being territorial if necessary, sleeping and being sociable if they wish.

Grooming

Another of those behaviours so strongly linked to cats is grooming. We take it for granted and reap the benefits of a clean, non-smelly pet living with us. We also know that grooming behaviours are important to a normal healthy cat and that a lack of them may be a hint that the cat is depressed or unwell. It is worth taking some time just to look at how cats groom, thinking about why they groom and understanding the different functions of grooming. Grooming involves both the tongue (which has backward-facing barbs, used in feeding but also vital for effective combing) and the teeth, which pull out burrs or tangles. You often see cats using their teeth to clean between their toes and chewing off bits of nail which are frayed or falling off.

Grooming for cleanliness

There is a huge motivation to groom for cats. Presumably grooming keeps the coat clean so that they do not carry a strong odour when they hunt by stealth, but remember that the coat also has very sensitive hairs which as they move give the cat information about what surrounds it. It is a valuable and sensitive commodity – if hairs are pulling or stuck together the information may be missing or false. Grooming also spreads oil from glands in the skin along the coat, which keeps it healthy and makes it waterproof; getting wet and then cold is not a good survival mechanism. Grooming also spreads the cat's own scent over the coat – grooming from the tail downwards along the legs will spread scent from anal glands that are important in scent communication. In hot weather cats don't pant like dogs but the saliva which they put on the hair and skin evaporates, cooling them.

Mother cats have to groom their kittens from birth, not just to keep the coat and skin clean but to stimulate them to urinate and defecate until they are old enough to be able to move away from the nest to do it themselves. This also keeps the nest clean.

Kittens learn to groom, rather awkwardly at first (and apparently if they are not cared for well as kittens, they may not groom themselves very well in later life), starting in the second week of life when the kitten tries to groom its front paws and then progresses to the rest of its body. By the time they are mature, cats may spend up to half of their waking time grooming, keeping the coat unmatted, removing old hair and dander, and removing fleas and parasites if the cat is not lucky enough to have an owner who treats them.

First the forepaw is given a good covering of saliva by licking several times and then it is wiped across the neck and the back of the head and ears and finally over the face. The cat uses circular movements, often using a ducking movement of its head and neck which allows this to happen slickly. The cat will then lick its paw again to replace the saliva. It's amazing how effective this is in keeping the head clean, but it definitely works! Cats then lick the front legs and move on to the back legs and tail. They usually groom symmetrically and systematically unless there is something urgent to be taken care of which has soiled the coat and needs to be removed. Grooming is most often undertaken after sleeping or resting, and after eating, when the cat licks around its mouth area. To reach all the parts of its body, the cat can manage some very unusual positions that make its continued suppleness necessary well into older age. Indeed, our super-senior cats may begin to show their age because their coat starts to look unkempt.

The need to groom is highlighted by research that showed that if cats are stopped from grooming for three days, they will increase grooming by almost 70 per cent in the next twelve hours. It is something to consider when cats have to wear those plastic collars to stop them getting at wounds – the desire to groom must be very strong and wearing a collar for long periods of time without relief to enable grooming must be quite stressful. Smaller and more flexible collars or periods when owners take the collar off and let the cat groom (while making sure they do not do any damage to healing wounds, of course) will be much appreciated by cats and help the healing process by reducing stress arising from not being able to groom.

Cats spend a huge amount of their time grooming and the instinct to clean the coat and paws even seems to override a cat's normal sense in avoiding poisonous materials which they would not normally eat or drink. Unfortunately, grooming means that something poisonous or caustic can be swallowed and ingested and may damage the tissues in the mouth to take the poison into the digestive system. Examples might be brushing against a newly painted fence and then grooming off toxic substances like creosote, or grooming the paws after walking through something poisonous such as antifreeze or ant-killer containing permethrin. Cat sensitivity to poisoning is something for cat owners to be aware of, for example there is often a flurry of poisonings in the spring after people scrub their patios with a cleaner that contains benzalkonium chloride but do not rinse it off properly. Cats walking through it before it dries, or when it rains and wets the patio again because it has not been rinsed, will find their paws coated and they will become ill when grooming their feet. It has been suggested that because cats groom environmental pollutants off their paws and coats, they could be useful in estimating how polluted an environment is – this makes sense where cats live on the streets in areas which are unused or derelict and may ingest dust that contains different substances which could be measured in hair samples.

Grooming for communication

When two cats are happy to be together, they often lick or groom or rub against each other. We call one cat licking another 'allogrooming', and it usually includes areas around the head and neck, the most difficult places for cats to access themselves. Cats rubbing their heads over the head or body of another cat or even rubbing along the cat's body is called 'allorubbing' and is thought to enable cats to share and exchange body scents and form a group odour.

Even though cats only groom each other if they get on, it seems it is usually cats that are more dominant or higher ranking which do the grooming of lower-ranking cats, just like a mother cat grooms her kittens. Behaviourists are loath to use the word 'dominate' or 'higher ranking' when referring to cats because they do not have a structured social structure in the way that dogs do. However, if you have several cats, you may think that one is rather less likely to back down than others or is confident about what it wants and getting it, or even bullies others to push its agenda forward! This is the one that does the grooming. As the groomed cat seems to thoroughly enjoy the grooming, it seems a strange way around, but perhaps it is a sign of acceptance between cats.

Grooming for other reasons?

If cats do something clumsy, like slipping off the back of the settee, or if they are startled or confused momentarily, they often stop and groom to sort of pull themselves together and almost to avoid embarrassment (that's the human speaking!). It can also relieve tension after an altercation with another cat. The grooming may seem rather out of place in the middle of conflict or in a moment of tension, but is thought to be a displacement activity and seems to calm the cat. Cats sometimes take grooming too far when they're feeling stressed and unable to find ways of coping and this may result in them breaking off hair and making bald patches.

If grooming is so vital to cats, it should make us very aware of its uses and of interfering with it. It also raises the question as to why we are producing pedigree cats with little or no fur – how does a tongue with barbs on it affect the skin if there is no hair to protect it? What happens to the oil produced by glands in the skin to be spread over the hair to make it waterproof?

What do bald cats do to calm themselves? How do they keep warm? Presumably they live totally indoors in warm houses, or their owners have an excuse to dress them in coats? A lack of hair will also mean a lack of whiskers or, often, distorted whiskers – is this a loss of one of the cat's major sense organs?

At the other end of the spectrum, why are we breeding cats with coats so long and with so much undercoat that they cannot sort out themselves? Breeds like Persians have coats so long that they probably find it hard to reach the end of the hairs as they groom, and the undercoat is so thick it is just too much to deal with. Does this type of coat matt more easily too? Is it too hot during the warmer months of the year? Cats have developed longer coats when they live in cold climates, but they still have to be able to be self-reliant and keep them clean and dry. If the undercoat is matted, then it is unlikely to be waterproof and in fact will just absorb the wet and constrict like wet wool, pulling the skin. A matted coat pulling on the skin and probably causing irritation to the skin must be torture for the cat. We may believe that owners will do this for cats with long coats and so it's OK, but how many owners get to a point with cats where grooming becomes a big threat to their relationship, the cat trying to avoid being groomed and the owner desperate to de-matt the cat? People often don't have the time or patience or knowledge to groom their cats in a painless and positive way. Grooming can become a battleground and the joy of living with the cat is diminished. Many end up being clipped at the veterinary clinic – what a shame for the self-reliant cat to have to suffer being shaved (and likely sedated as well in order to do it safely). A matted coat or a shaved coat removes all the benefits we have discussed of the coat. Should we be breeding cats that can't care for themselves?

Somewhere to toilet

We know cats are fastidious about their toilet habits – it is part of what makes them great pets. Mother cats teach their kittens very early on to urinate and defecate away from the nest or in a litter tray in our homes – kittens watch and learn, and digging and covering is instinctive. So, cats want places where they can dig a hole and cover it over; somewhere that is safe and accessible. And where that tray is placed is important – see Chapter 2. When cats start to urinate in places in the house where we do not want them to there is going to be a reason for that: Are we cleaning the tray often enough? Is it in a place where they feel worried about using it? Have we changed the type of litter they don't like? Are they ill or feeling threatened or anxious about something? They aren't doing this because they're being spiteful or dirty or naughty, and it's up to us to try and figure it out.

Claw sharpening

Our cats do a lot of claw stropping or sharpening inside our houses or outside. The claws are an amazing mechanism and much more complex than the human fingernail. The cat's claws attach to the end of the toe bone, fixed together by ligaments. Most of the time our cat's claws are not visible. When it is walking, resting or relaxed the sharp nails are tucked into a pocket of skin. This means the cat can walk silently and those sharp points do not get caught in anything or become blunted. Claws are normally sheathed until the cat wants to use them, when muscles and tendons in the foot pull the end bone of the toe forwards. This pushes the claw out and straightens the toe so the claw is rigid and strong – getting claws out rather than retracting them from an out position is a positive decision to make. The bases of the claws are well served with nerves that tell the cat how extended the claws are and about movement from side to side – the cat is very aware of what its claws are doing and they are very sensitive. They give the cat a great deal of information and we know how delicately they can be used, with people, for example, compared to their fierce use in hunting. Bearing all of this in mind when we talk about declawing cats brings home how horrific it is. The sensitive pads of the paw help the cat to move silently as well as picking up textures and vibrations.

And the claws/pads are multifunctional, as we have learned about lots of the other physical attributes of cats. The claws are used in hunting, climbing, defence and communication (scratches leave a visual mark, and glands on the feet produce scents which are spread over the surface during scratching). Once again, this behaviour is absolutely inbuilt so that the cat's weapons are kept in tip-top condition; therefore, they need somewhere to express this behaviour.

Cats that go outdoors may or may not have access to scratchable things like wooden posts or trees which allow them to grip with the claws to the right depth and then pull downwards to pull the outer covering off the claw to reveal a gleaming and very sharp new point. Cats that go outside may also scratch inside; there is no guarantee that they won't do this indoors if they go out. Those that don't have access to the outdoors therefore definitely need an outlet for scratching behaviour. Scratching is also linked to territorial marking – another highly driven behaviour in cats, so we should not underestimate the cat's need to undertake this behaviour. In many places in the world, homes and furniture are valued above cat welfare and the claws of cats are removed – more on this in Chapter 7.

Hunting and play

As we have seen in Chapter 1, cats have been designed by nature over millions of years to be fantastic hunters. Mother cats start to teach their kittens what hunting is all about very early on, bringing back injured prey when the kittens are about four weeks old so they can develop their skills.

The fact that cats will hunt whether hungry or not shows how inbuilt the behaviour is – if you are surviving on your wits and talents you cannot walk past an opportunity to gain food, and so movement will automatically trigger behaviours aimed at catching prey. Apparently only about 10 per cent of hunts are successful, which means that without human food, cats need to be hunting a great deal in order to have enough to survive. We benefit from this drive in our play behaviour with our cats – the rapid response to a dragged piece of string or a bird-like toy. With our pets we may find that they are not too interested in hunting or don't catch a great deal, or they may be keen hunters. Play is closely associated with hunting and is key to discovering and using the cat's senses and abilities to locate, catch and kill prey. Will lots of play replace the need to hunt? I am not sure we know but we like to think that using all of those talents in a play activity will mean that the cat has exercised, both mentally and physically, its talents in that area. If that is so, then we can help to reduce the hunting that very few owners want their cats to do and improve the cats' wellbeing with an outlet for their energy.

Play is also something we associate with cats, from playing with toys to playing with prey. It is difficult to separate behaviours that constitute play and which are predatory, as cats play in different ways when not hunting, both with other cats and with objects. We really don't know a great deal about how and why cats play, despite the fact that they do it every day in our homes! More about play in Chapter 8.

Sleeping

Cats are one of nature's best sleepers, from cat naps to deep sleep, depending on the circumstances. They may slumber for

60 per cent of their lives, which means they spend twice as long asleep as most other mammals. Lions will sleep a great deal after gulping down a carcass because the food will keep them ticking over for several days. While herbivores have to munch away all day on vegetation in order to meet their energy requirements, a feed of meat is rich in calories and nutrients. The carnivore must exert more energy to catch its meals, but it is also able to rest and digest between them. Our cats, being well fed, have a lot of spare time to sleep.

Cats often fall in with the daytime patterns of their owners, choosing to sleep perhaps when they are alone during the day and being active in the morning and evening when their owners are around. At weekends they return to regular periods of short catnaps when they take forty winks and have periods of deeper sleep safe in the knowledge that we are at hand if required.

When catnapping, cats will settle on any spot and close their eyes while remaining fairly alert. A napping cat will be well aware that you are around or that you are approaching because its ears will still be on 'radar patrol' scanning for any sound. The cat that is actually sound asleep may get quite a shock if awakened by a loud noise or sudden touch, so if we have to wake our cats, we need to do so gently and softly.

One German animal expert studied more than 400 sleeping cats and concluded that he could tell the temperature of the room by the position that cats took up when sleeping. At less than 55°F or 13°C, cats were curled up with their heads tightly tucked into the body but, as the temperature increased, the cat's body shape opened up. At over 70°F or 21°C, cats uncurled and had their legs and paws out in front of their bodies. Cats may end up with their feet in the air or lie on their side if they feel safe and warm. We may be able to recognise how our individual cats react to temperature and safety by the ways in which they lie.

Cats originate from warm parts of the world and seem to absolutely love being warm, either in the sun or somewhere cosy in the house. They are brilliant at finding the softest places too and making best use of human furniture or bedding. Cat sleep follows a much more fragmented pattern than human sleep – understandable if they have to hunt at the best times to find prey – often at dawn and dusk but also during the day. A confident cat may not worry too much about where it sleeps but a more nervous one may need to look for somewhere to feel safe before it relaxes and stops being on high alert. Kittens need more sleep, just like babies and older cats.

It will not be a surprise to hear that sleep is a need, and usually it is not a problem, but consider how we keep cats sometimes with other cats which they do not get on with. If in a confined space such as a flat from which they cannot go outside, it is very difficult to get away to a safe space. Bored cats may also use less confident cats and 'bully' them as an outlet for energy and activity, so the cats at the receiving end may not rest well.

My behaviour expert friends also mention 'feigned sleep',

which is something I have not really seen for myself but can be seen in cats in homing centres that are very stressed, anxious or frightened. While they may appear to be sleeping, they're not and they aren't relaxed – the body is tense and ears upright and taking in every sound. While their eyes may be closed, the cat is very aware of its surroundings – it almost seems as if it is closing its eyes to help it disappear and hoping that is not being noticed. If noticed, people need to consider ways they could reduce the cat's stress.

6.

What does liking cats say about us and what do cats think of us?

CATS ARE BECOMING more and more popular world-wide as pets – they are easier to keep than dogs and provide companionship to an ever-growing number of people, particularly people who live on their own. However, our lifestyles, attitudes and expectations are changing too, and more and more cats have to fit in with us. Where are we with how we are understanding and responding to our cats and making their lives with us healthy and happy? How can we understand our cats better, listen to what they are telling us in all the ways we have discussed, and then talk to them through all the mediums available to us – from providing an environment and ambience where they can be themselves, can express what behaviours they need to, and can lead us on the interactions they need, so that we can have a mutually enjoyable relationship?

Reading this you may think, 'This is all about what the cat needs', and indeed it is. We have control over their lives, and bringing cats into our homes requires adaptation from them, so we should be willing to adapt, too. For many people that adaptation may not be difficult, can actually be very enjoyable or they may not even notice that they are collaborating more because it comes naturally to them. For others, with cats that don't perhaps fit into what we 'expect' of our pet cats because they are more distant due to all the reasons we have discussed, that might be more difficult, but understanding and trying will undoubtedly help the situation. Respect for the cat and enjoyment of what is possible can lead to small steps in interacting with them, which can be very rewarding. If not focusing too much on your cat and trying to force it to enjoy stroking or cuddling allows it to come forward and volunteer interaction, then that should be recognised for the achievement it is.

What does liking cats say about you?

If you are reading this book, you are probably a cat lover or at least curious about cats and have bought a copy yourself or been given it as a present because you are recognised as liking cats. In a world where we like to put labels on everyone, what does this perhaps say about you? People also seem to be obsessed with pitching cats against dogs and cat lovers against dog lovers. Surveys are continually being carried out to see if there is a difference between 'dog people' and 'cat people' and it does lead to a lot of 'assumptions'. Interestingly, in the world of dating and judging profiles written to attract someone else, cat lovers may be the butt of jokes or unflattering descriptions – such as that people who don't like dogs shouldn't be trusted and, of course, the ultimate put-down for women who have multiple cats is being called 'crazy cat ladies'. One survey suggested that some women viewing dating profiles featuring men holding cats rated them as less masculine and thus less dateable than those who weren't. But perhaps others thought that men with cats showed more sensitivity and might be better partners. Research from the 1980s pointed to male cat lovers as being more autonomous (controlling their own lives) while male dog lovers ranked higher in dominance (power and influence over others and controlling a situation) and aggression. This may suggest that in the dog-owning situation dog obedience is important. Female cat lovers were lower in dominance and thus perhaps less worried about control and the nature of cats, more independent and definitely not centred around obedience. Females from both dog- and cat-loving groups were lower in aggression. However, a lot of research is taken from surveys based on veterinary visits or questioning dog and cat lovers (who may be somewhat biased) – and indeed we respondents have already been influenced by the stereotypes which are so liberally used.

An online survey carried out by the University of Texas at Austin asked people to rate themselves as 'cat people' or 'dog people'. Almost half of those who took the survey called themselves dog people, a quarter said they liked cats and dogs and only 12 per cent said they are cat people. The survey found that 'dog people' were 11 per cent more conscientious than 'cat people' – self-disciplined and having a strong sense of duty. They were also 15 per cent more extroverted than cat people in the survey – meaning they were more outgoing, enthusiastic, positive and energetic than the professed cat people; they were also 13 per cent more likely to be 'agreeable', meaning trusting, altruistic, kind, affectionate and sociable. But 'cat people' were 11 per cent more likely to be open, meaning curious, creative, artistic and not traditional thinkers, likely to try new things and spend time on their own. They found that cat people were 12 per cent more 'neurotic' than dog people in that they were more easily stressed or anxious or were worriers. However, as you can see, these differences are not huge at all and do differences of about 10 per cent make us think there are huge differences?

Other surveys looked at cat owners compared to non-pet owners (rather than dog owners). Findings in one showed that the cat owners had a lower level of mental health disturbance and therefore better psychological health than the non-pet owners. The cat owners also had a more favourable attitude towards pets (probably not surprisingly). No significant differences were found for depression, anxiety, sleep disturbance, ability to care for someone or a tendency for people to present themselves in a generally favourable way between cat owners and non-pet owners.

Other similar research has found that cat owners are more socially sensitive, trust other people more, and like other people more than people who don't own pets. Women were more likely to label themselves 'cat people' than were men. Another study suggested that dog people were warmer and livelier, more conscious of rules, and more socially bold. Cat people, on the other hand, were abstract thinkers, more intelligent and more self-reliant.

In 2016, in honour of International Cat Day, which takes place on 8 August every year, Facebook looked at the social characteristics of the two camps of dog and cat lovers. They referred to the stereotypical animal characteristic of dogs being more social and easy-going, while cats are reserved, independent and unpredictable, and asked if any of these characteristics were reflected in people in the US who shared pictures of dogs or cats or both on Facebook. Not surprisingly, cat people tended to be friends with other cat people and dog people with other dog people, but not exclusively so, and in fact cat people liked dog people too, i.e., liked other animal-friendly people.

About 30 per cent of cat people were single, compared to 24 per cent of dog people. But being single and a cat lover was actually quite inclusive! It wasn't just older women (a conclusion many might jump to) – younger cat lovers, and male cat lovers of all ages were just as likely to be single. Some of these relationship results may be due to urban/rural differences in where cat and dog people live, as dog people may be more concentrated in rural areas, where there's more space for a dog to exercise, while cat people were more often found in cities (though both groups lived in both city and country). Cat people disproportionately liked books, TV and movies and were especially fond of fantasy, sci-fi and anime, while dog people liked stories and things about dogs. In the US and the UK more households own dogs than cats, but there are more pet cats than pet dogs overall as there are multi-cat households and many which have both dogs and cats.

How do cats see us?

We've looked at cats themselves and how they communicate, what they seem to want and need alongside our own human needs, and the things we want from and for our cats. So now it's time to move on to look at what we know about how cats see us, how they interpret our signals, how they interact with people, what influences that and how their behaviour is changed by our presence. There is much more research carried out on dogs than cats and sometimes tests for cats are simply adapted from dog experiments, but as you can guess, this doesn't always work with cats.

Cats do not cooperate in such a straightforward way as

dogs, but as one researcher has said, 'If you're having trouble measuring behaviour in the species, it's probably not the species that's the problem – it's the methodology.' Often dogs have some training before they take part in tasks and food is frequently used to reward them. You can train cats, as my colleagues Sarah Ellis and Linda Ryan can easily illustrate. It requires thought and patience but is certainly achievable. Most of us are probably just too lazy to do it well. We do train our dogs, but I suspect they are keener to cooperate or at least show their cooperation and are more forgiving of poor technique than cats! Owners may say that their cats are attuned to their behaviour and respond to it, but despite their popularity as pets, little research has been done on cats' social relations with people.

Do cats prefer certain people?

A lot of the research that is available to us has been done by Dr Dennis Turner and his colleagues (who commented on the individuality of cats, which has influenced this book), beginning back in 1982 when there was little information available about cat behaviour, especially in relation to their interactions with people. Turner's comments that women were the most likely people in a household to interact with cats may have been because more women stayed at home rather than went out to work back then. Whether this is still the same now should be borne in mind, as roles change, genders become more fluid and more people work to achieve enough income to live – although following the pandemic many people of all genders also work from home now too. However, it probably still holds that women are the people in the home most likely to feed the cat and interact with it. Not surprisingly they found that women with children had less interaction time with cats, as did outdoor cats and individual cats in multiple-cat households – indoor cats had more interaction time, as did single cats. Research has also shown that single cats were more active at home, stayed closer to their owners, were more playful and interacted more often than cats in multiple cat households. This all makes sense.

Interestingly, cats do not spontaneously prefer one gender or age of person, but the different behaviours of the people in these groups causes cats to react differentially. The group found that the way in which people approach cats affects how cats reacted to them. Adults interacted more often while seated, men more often from a sitting position, while women and girls got down to the cat's level. Women talked to the cat more than men and the cat talked back. Children were likely to be standing when they interacted with the cat, and boys were more likely to go to the cat and follow it around (perhaps even chase it), while adults waited for the cat to make the first move. Older people were happy to wait for interaction and were often rewarded for their patience by cats spending more time with them.

Who starts the interaction?

Then came some findings that were more interesting but perhaps something cat lovers might not want to hear! When people initiated an interaction with the cat, the time spent in that interaction was less than if the cat made the first move towards the person, perhaps approaching with their tail up, rubbing around people, meowing or even gaining attention by scratching furniture. But there also seems to be some give and take from cats – if the owner is happy to interact when the cat initiates it, then it seems the cat may be more willing to interact when the owner wishes.

More recent research which looked at cats in homing centres found cats spent significantly more time with people who were paying attention to them than people who were ignoring them, and that cats often preferred to interact with humans than be attracted by food or toys. However, a homing/adoption/rescue centre is a very different situation to a domestic home where human company is often 'on tap'. Some very sociable cats in these centres may feel rather starved of human attention and may take the opportunity when it comes to interact with people. Our pet cats at home know that we are very likely to react very positively to their pleas for attention and so can pick and choose when they want that.

As we have said, cats like choice, including the choice of when to interact and for how long. Cats are individuals and will vary considerably in how friendly they are with us and how long they enjoy our attention, which of course also relies on how we respond to them and how overwhelming they may feel that attention is. So even if cats don't seem to be interested in interacting with us, it's worth letting them know that we would be happy to interact if they wish. We then need to take our cues from them.

Do human personalities affect cats?

Recent research has found that the personality of cat owners seems to be aligned with those of their cats. People were categorised for their agreeableness, conscientiousness, extroversion, neuroticism and openness. Results seem to suggest that there are links between people and cat personality.

Owners categorised with the trait of 'neuroticism' (those with a tendency towards anxiety, negativity and self-doubt) were more often linked with cats said to have 'behavioural problems', which they reported as including showing aggression, anxiety or fear, or stress-related behaviours, and also suffering from medical conditions or being overweight.

The researchers also found that cat owners who scored higher on 'extroversion' (those who seek out social stimulation and opportunities to engage with others, and often described as being full of life, energy and positivity) were more likely to have cats that were allowed to go outside and were more satisfied with their cats, which seemed to be easier to live with.

It is very difficult when you look at research such as this, which is hard enough to undertake in the first place, in that you can't tell what causes what. My gut reaction is that the behaviour of owners which is influenced by their personalities is affecting their cat's behaviour, rather than their cats having more difficult personalities. It seems unlikely, doesn't it, that people who scored higher on neuroticism had by chance taken on more difficult cats either accidentally or on purpose? Perhaps they did choose cats that already seemed to be nervous or anxious rather than bold or confident cats? Perhaps they wanted to help cats and take on the more difficult ones? Perhaps they chose more nervous cats thinking that they could care for them and 'bring them around'? If not, then it seems to point to human outlook on life affecting cats in rather a distressing way.

We talk about 'problem behaviour' in cats and what we mean by that is usually that cats are behaving in a way or doing something in a place we do not like. This could be urinating or defecating outside of a litter tray in the home, reacting to interactions with people in a way which is seen as 'aggressive' but may just be a way of keeping people at a distance if the cats feel overwhelmed or do not want the attention. If we are rather more neurotic, do we just see things more negatively, and in fact the more positive people just didn't notice or care about some of the traits which the neurotic people seem to have experienced? If you are more worried about life, do you see more danger for your cat and try to control and protect it so much so that it makes the cat's life difficult? We have seen that cats like to take the initiative and like to have control over what they do. There is probably far more to think about on the human behaviour side than on the cat behaviour side!

Do cats bond with us?

According to the Human Animal Bond Research Institute, the human-animal bond is a mutually beneficial and dynamic relationship between people and animals that is influenced by behaviours essential to the health and wellbeing of both. This includes, among other things, emotional, psychological and physical interactions of people, animals and the environment.

More research has shown that cats can bond with their humans as strongly as we think dogs do and that we have underestimated this in the past. Most of us would instinctively say that our cats are bonded to us, but how do we prove it? Dogs may be motivated to bond with people because, like people, they absolutely need social connection. Cats are 'socially flexible' and very individual in what they do and what they like. Cats seem to see their owners and home as a secure place for them while things outside their territory can be threatening. We think it goes further than just a feeling of being safe and that cats and people bond because we like each other. Cats just don't express themselves in the same ways as dogs, or people for that matter. Understanding how we can develop these bonds to make our cats feel even more secure may help to improve our relationships with them or help them to deal with stressful situations, which could lead to them being more relaxed with us and interacting more. It would also help to improve the image of cats with people who for some reason don't like cats and need a reason for that dislike, such as cats being too independent or only being with people because they get something out of it.

The look of love

People are very sight orientated and we notice how we hold eye contact or look away – 'looks' are important. Eye contact also means something in the animal world. Researchers talk about 'gaze', by which they seem to mean to look steadily and intently and contemplate each other. Gaze is also a non-verbal communication between people and dogs and cats, and has meaning to the giver and the receiver. It also seems to contribute to the formation of a bond between the two parties. Both dogs and cats use their sense of smell, hearing, touch and eyesight to communicate with humans, but the species differ because they are very different animals, both in their behaviour with their own species and with people (because of how they have evolved and developed to live with and around people). Where 'gaze' fits on the spectrum that runs from a glance to a stare I am not quite sure, presumably it is not as aggressive as staring, but a bit more than just a quick look at someone. We know that staring can be quite an aggressive behaviour often used between cats when they are being combative. This may be different to the way dogs use eye interaction with us and each other.

There is a lot more known about the 'gaze' communication of dogs – their wolf ancestors live and hunt in groups, so looking at each other to communicate and cooperate is not difficult to appreciate. We know that dogs are affected when people look at them – apparently, they are more obedient when we are watching them, and may upscale their attention-seeking behaviour such as whining, whimpering or looking at their owners if their owners are looking at them. During their domestication dogs were, and still are, more under human control and reliant on their human carers, so perhaps being able to know who is likely to feed you because they are looking at you, or who will react to you looking at them, is very important. Dogs also follow a human's gaze when they look at something and may move their eyes between their owners and an object, for example when they can see but not get to food, perhaps to ask for cooperation with a problem. Dogs may rely on people to help them in situations where they do not know what to do or cannot solve problems themselves. The have evolved from a cooperative background, so are likely to look outwards for support because it has been a successful part of their development.

What about cats? We know that the wildcat ancestors of

our pet cats lived and hunted alone, so they did not need to interpret the gaze of others for cooperative hunting. During 'domestication' cats did develop the ability to live in groups if it suited them and if there was enough food, and where female cats cooperated in raising kittens. However, they don't have the range of behaviours which dogs use to signal submission during disputes, and do not align within hierarchical structures that help to guide and control the behaviour of a cooperative group and keep them together, something vital to social animals such as dogs.

For cats, if they do live together, the groups' 'glue' is developed by exchanging scents and by rubbing and grooming each other. Benefits of this group living may be having more cats around to deter other cats from coming in to use their food source and shelter, and to protect against predators or males that may kill kittens which are not their own. Perhaps cat behaviour was well formed in cats during their solitary phase. Mind you, we know that kittens learn fast by watching others, but whether this extends to noticing what their mother is looking at has not been studied. Like dogs, they can go to an object a person is pointing or looking at, but are less reactive to people's behaviours than dogs when confused or worried about the object – dogs seem to be able to ask for more help. I wonder too whether if the situation occurred in the cat's home with its owner, a cat might interact more because it knows its owner will react and may know that it wants something. When cats come across new or strange objects and do not know what to do, they seem to be able to read the person's facial emotional expression too. As cats did not evolve to gaze at other cats for food, this may have developed during the time they have been living with people.

However, a little bit of research has been done which shows that cats do show head movements and may change their behaviour because a person is gazing at them and seem to avoid a familiar gaze. Cats probably see a human gaze as the same thing as a cat's gaze (whether they see a gaze as a stare we probably don't know) and it wouldn't be difficult to believe as a stare may indicate a threat.

Dogs and cats can distinguish between signals based on people's emotions. Dogs can identify emotional states and facial expressions of people and change their behaviour depending on people's emotions. In one study, dogs sniffed, nuzzled and licked a human who was pretending to cry and dogs approached a strange object when their owners reacted positively to it and moved away from it when their owners reacted negatively. One study found that cats rubbed more against their owners when their owners were depressed, and that cat behaviour could distinguish between facial expression and postures. So, both dogs and cats can react to emotion – in dogs we would equate this to bonding with owners, so why not in cats?

What about the mood of people – does it affect cats and do cats affect human mood? It seems that women who own cats are less depressed and less introverted than women who haven't got cats. Cats seem to react to their owner's moods, but those moods don't seem to affect how the cat reacts. Women who are feeling in a negative mood seem to feel better interacting with cats, but cats do not improve moods in women who are already in a good mood. Single men and women were affected in the same way by cats, but men were more affected by the presence of women than of cats!

Research has also shown (and most owners could tell you) that dogs behave differently towards their owners than to someone they don't know, and their owners seem to be regarded as something which is secure – a bit like how children see their parents. Interestingly, when dogs and people gaze at each other, the hormone oxytocin (which is associated with birth and lactation and presumably with bond formation between mother and baby after birth) is secreted. Dogs looking at people caused an increase in oxytocin in people, which in turn increased the owner's interaction with the dog. Owner interaction increased oxytocin secretion in dogs too; in other words, both dog and person looking at each other saw an increase in the hormone, just like mothers and babies. Research has pointed to bonds between cats and people being similar to that between humans and dogs or babies. What we don't know is whether the gaze which helps the person/dog, person/person bond is the same for person/cat. My cats use gaze very differently with me, so once again it may be very individual – more on this in Chapter 9.

There is an interesting additional component of the

eye-to-eye communication between people and cats, called the 'slow-blink'. There are some new ideas about the relevance of the blink of an eye. Blinking in people can be done consciously but mostly we do it without thinking. In animals it is an unconscious signal.

When dogs and people gaze at each other, person and dog synchronise eyeblinks, with the dog blinking about one second after the owner, and owners blinking immediately after the dogs. The same phenomenon has been seen in cats and it is thought to aid mutual understanding and trust. Whether the blink breaks what could potentially be seen as a stare in cats may have that additional meaning for our feline friends.

Research found that when people slow-blinked at their cats, the cats often slow-blinked in return and that it can help owners to show affection to their cats. The same study also found that cats were more likely to approach strangers when they slow-blinked, perhaps enabling them to convey they were not a threat.

A slow-blink starts with a cat having a neutral expression. It then lowers its upper eyelids partially into a half-blink position. The cat then closes and narrows its eyes before slowly opening them again. Researchers think that cats slow-blink when they are happy, and the blink is a sign of trust and affection. It's very different to the highly focused look cats have if they are watching something anxiously or are feeling wary; the slow-blink is soft and relaxed and shows trust in the person.

Other things that help the bond

It makes sense that not only vision contributes to the formation of bonds between pets and people. For cats, their acute sense of hearing and amazing sense of smell may also contribute. We know that cats adapt their vocal communication to connect with people using the meow, which is generally only used for communication between kittens and their mothers. We also know that after being separated from their owners for a long period cats purr more than if separated for a short time. Cats can tell the voice of their owners and show more reaction with head and ear movements than when they hear a stranger's voice.

When cats live in groups, they tell the difference between members of their group and strange cats by smell. They have highly developed communication through smell. When cats live in friendly groups, they rub as they pass each other and leave scents on things around their territory. Scents that have been deposited build up a group scent with a particular profile that is reassuring to them and members of the group. Our cats rub against us and sit on our beds/chairs/clothes where we share scents (even though we are oblivious to this), so a reassuring smell can help them to feel you are part of their group, too, or that they are part of yours.

Dogs and cats have adapted their communications in different ways to help them live with people and create that 'mutually beneficial and dynamic relationship' that we call a bond. The phrase 'mutually beneficial' is one to think about. Very often when the bond is discussed it is from the human point of view – what pets can do for people, such as lower blood pressure, reduce stress and reduce loneliness or depression. By interacting with companion animals, elderly people can experience positive mental and physical effects. Similar results can occur in children during emotional, cognitive, social and behavioural development. But what about what the pet gets? Food, water, safety and healthcare are obvious, but sometimes we need to take the cat's position.

Misunderstanding our cats (either through ignorance or because we don't want to change our ways if we find out something we don't like) can mean we don't pick up on the cues that can tell us what our cats really think about being with us. It can be that we miss the positive and the negative messages they are sending, but recognising and responding to either can help us to live a better life together.

7.

Do we use cats?

THERE ARE THINGS I feel strongly about in our relationships with cats where I believe cats get rather a raw deal and their welfare could be called into question. In the next chapter we can celebrate the fun ways we live with our cats, but here, I just wanted to think about some of the situations where we ought to recognise that we can love them at the same time as misunderstanding them, and that we should sometimes question our intentions in order to make sure that we actually are doing the best we can for them. I am aware that this chapter may make some people uncomfortable or angry because they feel there is implied criticism, but working for cats for over thirty years has highlighted some areas in which I feel we need to understand our motivations and be more knowledgeable, even if it is uncomfortable. Exploring myself, I know that I have cats because they fascinate me, that I love having them around, that they make my home more welcoming and they are great company, but cats had no say in this matter! We have to be honest with ourselves about our needs and wants from and for our cats and what they actually need and want, too. From there we can truly try and do our best for them.

I was asked to give a talk at the World Horse Welfare conference – their excellent chief executive Roly Owers and I have known each other for a number of years and his conference really takes a forward-thinking approach to how we work with horses and examines the person/horse relationship. It's great to share ideas and challenge ideas across different species. He invited me to do a small spot about International Cat Care's Cat-Friendly Principles and asked if there was an approach we could take for cats which fitted with the theme of the conference: 'When does use become abuse?' It was an interesting idea and really got me thinking.

Of course, we don't consider our relationship with cats as 'use' and most people would be outraged to think that we do – we give love and try to help cats. We would probably agree that we 'use' farm animals to produce meat, eggs or milk, but cats don't fit this scenario at all. Horses perhaps fit in between farm and pet animals – they are used for work and competition and for people to ride for leisure, but they are loved individually as well. It got me thinking about how we as people associate with animals, what assumptions we make and how perhaps sometimes we cause harm without meaning to, without understanding or without caring that harm is being done because of our own desires.

Think about the different ways in which we live and work with cats: we have them as companions and also work with them as veterinary professionals, in catteries, caring for unowned cats, in grooming and breeding businesses, and by selling cat products or food. Pet owners want cats for companionship or something to care for, love or help. We have to enable cats to become pets, as we outlined in Chapter 2, so if we want to live with them feeling comfortable with us and behaving in a way in which we enjoy their company, we have to prepare them properly. The making of a successful pet (mutual enjoyment by cat and person) requires genetics for friendliness, positive human interactions in the first eight weeks of life, ongoing positive interactions, as well as good physical health and mental wellbeing. We tend towards anthropomorphism – attributing human thoughts and feelings to animals can improve empathy but may also detract from prioritising the needs of cats.

There are several areas where our partnership with cats can tend towards 'use' rather than being mutual, and where we really need to question our motivations and the welfare of cats in these situations. Here we will briefly look at declawing cats, breeding for extreme results and despite inherited problems, when our help can turn into a bad situation for cats, and how we might push them too far if we rely on them as emotional support animals.

Declawing

In many places in the world, homes and furniture are valued above cat welfare and the claws of cats are removed to prevent them causing damage by scratching. This procedure started around the 1950s and became popular by the 1970s. Many countries realised that it was a horrific thing to do to a cat (more details below) and it is now banned in over forty countries. It is still allowed in North America, but some cities (New York is the first state to ban it) and some Canadian provinces now also ban the operation; however, an estimated 25 per cent of the feline population is declawed. If you look up declawing on the internet one of the shocking pictures you will see is of a pile of little claws which have just been removed from a cat's feet.

So why is it so horrendous – isn't it just like cutting the nails? The truth is, it is much more extreme. Declawing is actually the amputation of toes – think about removing your fingernails by cutting off the top part of your finger from the top joint at which it bends (with a type of nail trimmer, a laser or a scalpel blade) – that is declawing. If the surgery is not performed well the tissues are damaged and small bits of bone can be left behind. The wounds may also become infected, or pieces of nail may grow back. Even if it is well done, we know that removal can lead to long-term pain and behavioural changes such as urinating and defecating in the house, hiding or behaving in ways which push people away (probably thought of as aggressive behaviour) and chewing fur – not signs of a relaxed and happy cat. There may also be damage to the nervous system and the cat may suffer from pain similar to that suffered by people who have limbs amputated.

As an alternative to removing the claws altogether, an operation which prevents the cat from retracting its claws by cutting the tendons (which affect the cat's ability to pull the claws back into their sheaths) can also be carried out. It allows the cat to keep its claws but prevents it from scratching to remove the outer sheaths of the nail. This means that claws are not used properly, can become overgrown and will need constant clipping – more need for handling the cat, which may not be welcomed by many. Also, remembering the strength of the cat's instincts and drives to carry out these behaviours (which would be the difference between survival or not for them in a world outside that of people), how will a cat feel when it cannot actually carry them out? Removing claws or preventing them from being used will not remove the need for cats to scratch – they cannot overcome an inbuilt instinct.

Declawing is a painful procedure and can have short-term and long-term effects. Cats may not want to put weight on the declawed feet or to move around. Research has also shown that cats may develop back pain and that declawed cats have significantly more problems than non-declawed cats. We do not think of the supple and agile cat as having back pain, but if a cat is trying not to put its weight on all or some of its paws, its gait is likely to be changed and this may then cause back problems.

Declawing has been argued as being pro-cat because it stops people euthanising or abandoning cats because they cause damage to the home. What kind of an argument is that? Indeed, declawing may actually mean that they are uncomfortable or in pain and this may well increase the chances of problem behaviours (seen by owners as problem behaviours), which may ultimately lead to rehoming or euthanasia for many cats.

There are ways of reducing damage indoors caused by scratching by providing scratch posts and other materials which are the right size, shape and placement to encourage cats to use them and to encourage and reward them for doing so. Cats with access outdoors may not scratch so much indoors and claw trimming may reduce damage, however it does require close handling which may not be popular with some cats. It may not be easy to prevent or control, but perhaps we as owners must accept some imperfections in our homes to accommodate cats and their behaviours. Scratching is a need which has to have an outlet. Removing the claws, in my opinion, is definitely an abuse. For many people it may be seen as quite a normal thing to do because it is so common, but they may well not be aware of the true horror of what they are doing to their pets.

Pedigree breeds likely to have problems

Nobody, not even someone who doesn't like cats, could argue that a cat is not a thing of beauty. Lithe, pretty, with such beautiful coats and eye colours, the whole cat family has evolved to be top-of-the-chain predators unaware of this beauty, which is something that is also very appealing to people. Many of us appreciate seeing our cats every day just because they are so beautiful. You may think that this is perfection, and we should be happy to appreciate it. However, we are human and our tendency is to try and have more of what we like and to change it if we can.

We have questioned the use of the word 'domestication' with regard to cats as we feel they may not be as domesticated as many of our other pets. Most animals when they are 'domesticated' have no choice over the mate they produce offspring with, but probably over 90 per cent of cats in the world are non-pedigree, having a random mix of genes. We seldom choose the mates for our non-pedigree or moggie female cats because mating happens outside without us seeing; in fact, we often don't even know what the male looks like, let alone his character – the female cat decides. When we breed pedigree cats, we do control the cat's genetic makeup by choosing to breed from cats with certain traits – usually physical traits to make them appeal to people or to create something new.

What is a pedigree cat?

A pedigree cat is one where we can pretty well predict what it is going to look like, and it often follows a 'standard' laid down by organisations that register pedigree pet cats and are involved in putting on cat shows that reward owners of cats who conform best with this standard. This ensures that cats of a certain breed look similar, albeit perhaps with different colours of coat or eyes, which are allowed within the standard.

Pedigree cats, like most domesticated animals such as cattle, sheep or pigs, don't choose their own mates – people select cats to put together, primarily for a certain look. Good breeders will choose their breeding cats according to health, temperament as well as 'look', but this does not always happen. Hopefully they also choose cats because they are friendly and will be happy pets. Pedigrees are bred from specific and very similar others, all from the same 'breed' or with 'allowed' other breeds either within a certain breed standard or to create a new breed. However, this also means that the number of cats available to breed from is much-reduced and hence the 'gene pool' is not large. It is known that the smaller the pool, the greater the risk that any problems caused by genes in that pool will be spread to those cats.

So, while the look of the cat you are seeking can be pretty much guaranteed, problems can arise because of the limited gene pool. If there are health problems or disorders within this pool, these can be passed on more readily to the next generation – this is what we call 'inherited disorders'.

We have also taken some breeds and pushed their physical characteristics to extremes, examples being very flat faces, short legs, lack of or very long coats. Others have had physical characteristics selected and bred for that in fact cause pain and discomfort for the cat, such as Scottish Folds (see below).

The 'look' or 'standard' of some breeds may require what could be considered as extremes of body form – such as very short noses or lack of a tail, which may not be wonderful in terms of the cat's health. Let's look at three breeds whose physical characteristics are no doubt causing problems for the cats born into those bodies which have been created by people.

Scottish Fold: The Scottish Fold is a cat with a pretty normal body and head shape but with ears that fold down and forwards, like the flap of a birthday card envelope. It gives the cat a very cute look with a round face and big eyes that is appealing to people. So what is wrong with that? Again, it could be argued simply that cats need ears that stand up in the normal way in order to hear properly and to be able to communicate and that should be enough. However, that is not the biggest issue – the ears are affected by a gene mutation which affects cartilage, causing a weakness that means that the ears do not stand up straight but fold over. Unfortunately, and not surprisingly, it is not just the ears that are affected: cartilage in the joints, which is needed to protect joints and enable them to work properly, is also affected and this leads to damage and the development of arthritis. If the ear is affected, then so is the rest of the body. So, for that cute look there is a very high risk of pain for these cats from an early age. This is a high price to pay for 'cute'. Being cats, they will not limp, cry or show pain, but may not move around much and suffer silently – the X-rays and veterinary examination will tell a very different story as to what is happening in those joints. Visit the International Cat Care website to see for yourself.

The Manx cat: The short or absent tail of the Manx cat which defines the breed is caused by a mutation that affects the spine. There are in fact different tail lengths from normal to taillessness, with names like stumpy and rumpy to describe the degree of shortness. What is the problem of having no tail? You could argue that a cat has evolved to have a tail and therefore it is there for a reason. The tail aids balance and is used in communication, which are good enough reasons not to aim for a tailless cat.

However, the additional problem is that the mutation causing the loss or shortening of the tail in the Manx affects not just those issues, but also the spine and spinal cord and the nerves. This results in a form of spina bifida – a developmental abnormality of the spine that can result in problems with control of urination and defecation, and sometimes also with control of the back legs. How has it become the symbol of the Isle of Man? Presumably at one point the tailless cat was common because the gene responsible for the lack of tail is a dominant gene and only one copy of it is needed to produce the defect – only one cat in a mating needs to have the gene to produce kittens with a short or no tail at all. In fact, the gene is so lethal that if you mate two cats with no tails the kittens are likely to die before they are born. Even when a tail or part of a tail is present, the problem may not be absent, and vertebrae will still be fused. Some Manx cats have also been reported to develop severe and painful arthritis.

At the time when the cat was recognised as special to the Isle of Man and believed to be a good symbol for it, the cause of the taillessness wouldn't have been known and the problems not understood. However, surely it is now time to reassess whether we should be continuing to breed these problems on purpose and to use changes made as an example of good welfare for animals.

Persian or Exotic breeds: The third example is not a single problem but a question of how far we are willing to go to achieve the looks which, for some reason, we are aiming for. The Persian and Exotic (essentially a short-haired Persian) have very flat faces – what we call brachycephalic. In years gone by the Persian had a normal face and the nose was not so flat but, over time, breeding from cats with flatter and flatter faces reached the point where the nose is now at the same level as the eyes and the face is almost concave. There are people who would like these breeds to be less extreme and this is a matter of degree, which in theory could be reversed, but there is no doubt that changing the skull shape has a severe effect.

It is the same issue we are seeing in dogs such as bulldogs and pugs and comes with similar problems associated with the changes, or perhaps we should say distortion, in skull size and shape. If you took a normal cat skull made of soft clay in your hand facing forward and, using your fingers, pushed up the lower jaw while pushing the nose area inwards, the whole skull twists and changes shape. The tissues and structures (which have very specific functions) inside the skull have to go somewhere else or will be squashed – this is what happens as the cats are bred to extremes. The jaw and teeth move and may not function properly because the teeth are no longer aligned in the upper and lower jaws. Many of the pictures of cats on the internet that are considered funny because they are ugly are of cats with teeth that stick out at strange angles or jaws that are aligned in a way that makes them look angry or just weird. This is going to affect eating and grooming.

Poor Persians – not only is their extra-long coat and thick undercoat unmanageable, but the fluids which are normally produced to keep the eye moist are not being spread over the entire bulging eye (pushed out and more exposed because the eye socket is shallower), which can lead to eyes becoming dry and painful. We know how painful it is to have dry eyes ourselves, and dry eyes are more likely to be damaged and ulcers arise, which are no doubt extremely painful. The hydrating and lubricating fluids are still produced by the glands, but the tubes that normally carry the fluid away after it has kept the eye moist now overspill onto the face – because the tubes are distorted by the twisting of the skull to make it into that flat shape. You will see stained hair on the face of many flat-faced cats and the dampness can also lead to skin infections in folds on the face. Think of a moggie cat – anything on its coat or face will be cleaned off immediately – it can feel the fur becoming sticky or wet and deals with it as soon as possible (remember in Chapter 1 we looked at how sensitive the cat's coat is).

Add to this that the jaw of brachycephalic cats is often malformed because of the changes to the shape of the skull, and grooming may be more difficult (even eating can be affected). Persians are reported to be quite placid cats, which may indeed stem from their breeding, but I can't help thinking that in some way they have to cut out some of the normal feedback they have from the sensors in the skin which tell them the coat is wet or tangled. They have to put up with people wiping faces and eyes. I think of my children's reactions to having their faces wiped when they were small – they were not impressed and tried to duck and wriggle away to avoid it. I can image cats disliking it in a similar way! Not all cats will find wiping and grooming a challenge – some may enjoy it, but I know what my cats would think of it! When we have control of how a cat looks and this affects how it feels, surely we want to make that process as comfortable as possible? Successful grooming is needed and part of a cat's self-sufficiency – we should acknowledge and accept this and make it easy for the cats we produce to be able to achieve this themselves.

These problems are physical problems brought about because we want extremes and novelty, but there are also inherited problems that affect health and which are not visible from the outside. A small gene pool can mean that inherited defects like heart problems, kidney problems and other diseases which act to cause ill health can be easily shared with all the cats. Because many of the cats are closely related any inherited disease can spread within the breed and if it doesn't show up obviously before kittens are bred from themselves, it spreads further. Examples of inherited diseases which cannot be seen from the outside are polycystic kidney disease and hypertrophic cardiomyopathy, among many. Good breed groups will notice, investigate and try to solve these problems by changing breeding policies and working with scientists to develop tests and then work to prevent them if possible (and it is not always easy if the mode of inheritance is complex). However, unfortunately some people are not willing to compromise their lines of cats which they have developed over many years because a problem has crept in, and you can have some sympathy for that, but cat welfare must take priority.

Should we keep all breeds?

There is something about making a breed which makes people then think it has a special place, has to be protected and must not be removed. Many of the reasons for not removing some breeds is that they are traditional or have become representative of their area or country and cannot be 'lost'. Researchers have looked at the genetics of breeds and, among the around-fifty recognised cat breeds, only sixteen are thought to be breeds that occurred in specific countries or regions because the cats there bred together and had a similar look – they had history of that place. The other breeds have been developed only in the past fifty years or so.

The look of some cats originally developed in an isolated place where certain characteristics became common to all the cats as they bred with each other – an example of a naturally occurring limited pool of genes. Most of these 'breeds' have then been changed and developed further by breeding and may be quite far from the original now. For example, the Siamese and Angora breeds developed in isolation from other groups of cats to form the basis of the cats we know today, but we have introduced further changes through selective breeding in order to produce colour varieties and changes in body shape. Another example of this is the Manx cat from the Isle of Man, which developed to have little or no tail (see pages 194–5). The Maine Coon and Norwegian Forest cat are the natural result of a mixture of cats which, by various means, reached the East Coast of the US and Norway. They grew long coats because of the weather, but this has been developed and these cats made into a breed with controlled breeding.

Some breeds were developed and originated by taking a cat which had accidentally been born with a genetic abnormality (such as short legs, folded ears or a lack of coat) and breeding it to create a new breed. Examples are the Rex breeds with their sparse curly coats, the hairless Sphynx and the short-legged Munchkin. If that 'new' characteristic causes no harm, then we have nothing to comment on, but if that alteration comes with health problems or makes the cat's life less easy, then we should think again. Should these mutations be pounced upon and turned into 'breeds' just because they are different?

Another way of making a breed is to take several other breeds and their specific characteristic and mix them together to produce a new breed with a new name. Breeds such as the Somali have arisen by cross-breeding, in this case by introducing a gene for the long hair to the Abyssinian. This seems to have been done carefully and with knowledge and has not brought any problems. Again, this can be done in a good way, carefully choosing the breeds to use which don't themselves have an issue, but it is often carried out using breeds with more 'interesting' characteristics such as hairlessness or short legs, etc., all of which may come with issues themselves.

Good breeders understand the breeds they are involved with and try to get to grips with the complex genetics to try and ensure they are healthy. Unfortunately, some people just take the external characteristics without understanding how they will mix and what they might mean to the cats themselves, which then turn out to have a combination of characteristics that may bring many problems. Examples of this include things like the Dwelf, which combines the short legs of the Munchkin, the hairlessness of the Sphynx and the curled ears of the American Curl.

The names of many breeds belie their origins. We place a lot of store in the history and origins of the breeds we back and prefer, and much is said in defence of not making changes, of losing a breed from a specific place or for historical reasons. Recently researchers looking at cat DNA from around the world have found that the cats generally come from four areas – regions of Europe, the Mediterranean basin, East Africa and Asia. They discovered that non-pedigree or moggie cats in the Americas were similar to moggies from western Europe. They also found that some American breeds, such as the Maine Coon and American shorthair, were genetically similar to western European breeds, suggesting these were descendants of cats brought to the Americas by European settlers. Interestingly, they also found that Persians, named for an Asian country, were actually more closely associated with moggies from western Europe than the more exotic name suggests. So the geographical identity of cat breed names may have very little to do with what their genes actually say and shouldn't be a reason not to question welfare.

Questions about welfare

For some reason there seems to be very little questioning about the sizes, shapes and styles of animals which are bred for us to have as pets. In this world where many things are challenged, the production of certain pedigree cats and dogs seems to be accepted or even welcomed for its novelty. Is this because we see breeders as creators and that breeds almost immediately become defended because they are defined or named, or because we like what has been produced and therefore are unlikely to question the welfare implications?

There are many responsible breeders producing cats that are both physically and mentally healthy and are suited to being pets and, as we will see later, they may even be able to improve how cats live with us by choosing cats that are less stressed around people and giving them the best start in life. As we've seen, there are cats with amazing coat and eye colours that are truly beautiful and healthy. If, however, there is even a small chance that the changes (often caused by a genetic mutation) compromise health or welfare, then questions should be asked, and cat lovers should demand that the new 'breed' is not proliferated. Both direct changes and those that arise accidentally need to be considered in the continued health and welfare of a particular set of cats which constitute a 'breed'.

Breeders of pedigree cats must accept responsibility for the production of cats with certain characteristics; purchasers of cats must realise that they create a demand that will then be fulfilled – both must put the welfare of cats before the rewards of money or creating or owning something new or different. The rule must be 'First do no harm' when we are altering what, it's not hard to admit, is a pretty perfect design in the first place.

Hybrid cats

We have the loveliest coat colours and patterns in our domestic cats and there is quite a lot of expertise among breeders to create and develop these. But sometimes even this is not enough. Some people want to go further and incorporate coat characteristics from wildcats. There is probably also a desire to create a different type of cat to create and satisfy a curiosity to live with a wilder type of animal. To do this they need to breed wildcats with domestic cats – I can see you raising your eyebrows already! Is that even possible? The answer is, in some cases, yes, although it is definitely not without issues. Several types of wildcat have been used in this type of breeding of what we call hybrids. The Bengal breed results from bringing together our domestic cats with Asian Leopard cats, which are about the same size as our pet cats but much shyer, being a solitary creature and probably highly territorial. The resulting breed seems to behave well with people, is bright and curious, but some individuals have a reputation for being very territorial and aggressive towards other cats in the area, sometimes terrorising and injuring them. This has happened enough times to give Bengals this reputation.

The Savannah is the result of mating between domestic cats and the rather larger wildcat, the Serval. There are concerns for the welfare of the domestic cats used in these matings – it seems to be veiled in secrecy but there are reports of injury or even deaths of the female cats used to mate with male Servals.

Gestation times also differ between the wild and domestic cats, which may affect kittens and the mother. Even when mating our domestic cats breeders have to introduce cats carefully and make sure there are no injuries – this is further exacerbated when you use a different species of cat and when the male wildcat may be a lot bigger than the female domestic cat, there is potential for disaster. There is also concern for the welfare of the wildcat males and the F1 crosses (the kittens from the first mating), which must be kept in captivity.

In the UK these cats need a Dangerous Wild Animals Act licence, but that is all about safety of people and does not cover the welfare of the cats and how they are kept. There is also concern for both the new owners and the hybrid cats of further generations which can be sold as pets because of uncertainty of behaviour and needs of the hybrid cats themselves. Other wildcat breeds are also being tried to create new hybrids in our desire for something different or to be the first person to 'create' a new breed of cat.

There will always be a demand for pedigree animals – we like to have a choice to buy and to align ourselves with, and some (not all) breeders like to make something new or different and want to adapt and change these animals to suit their preferences.

If, despite the best intentions of good breeders and owners, a problem becomes apparent within the breed that was not evident initially, the custodians of that breed should make any changes necessary to bring it back to health, including changing the breed standard or introducing new lines of cats (with appropriate advice), without delay.

If these changes come with no harm to the cat, then there is no problem and, as we have said, done well it may result in friendlier cats which are less stressed being in our homes and living alongside people.

I have mixed feelings about breeding cats – many years ago I had pedigree Siamese cats and really enjoyed their interactive personalities and I know many people who love their pedigree cats. But what I am not so fond of is the feeling that pedigree cats are somehow better than other non-pedigree, randomly bred or moggie cats – often just because their owners have paid a large sum of money for them. We should not consider any cat better than another – simply different. Just because we pay quite a lot for these pedigree kittens doesn't mean that they are any more important or better than moggies.

When the best of intentions can go wrong

There is massive work to be done to help unowned cats, and people working at the sharp end of finding solutions for these cats are to be admired – it is not easy; the need is great and resources of time and money severely limited. It is also very complex. As we saw in Chapter 2, a cat is not one particular thing and different solutions will be required to suit different cats. Done well, so many cats can be helped to have a happier and healthier life, some living with people. But as we have already admitted, human motivations can be complex, can be based on ignorance, in denial and driven by human need rather than cat need. Helping unowned cats can easily spiral out of control unless you are really pragmatic and realistic – the problems are bottomless, and resources are not.

Having developed thoughts on this area with my colleagues when I worked at International Cat Care, I understand how complicated it can be. The challenges are massive and while many think you are just picking up cats without homes and helping them along to new homes, it is far more nuanced than that. It can start simply enough but, unless controlled, can result in cats being 'rescued' and then kept in confinement in cages for a long time, without an understanding of whether those particular cats will even be able to live with people, while still looking for traditional homes for them. This situation can result in too many cats to care for and may lead to hoarding situations where the number of cats overwhelm people trying to help them in the first place and may lead to poor care or extreme neglect. Keeping a cat long term in a cage because it is perceived as safer for the cat than being on the street is not improving that cat's welfare.

The idea of homing centres or adoption centres lies in the title – it is to get cats out to new homes. If there were only a few cats to home, perhaps waiting until the perfect home comes along may be feasible, but such a luxurious situation seldom happens and a pragmatic decision must be made, one not based on personal preferences or prejudices but rather whether the cat would be better off out of a crowded environment, which may lead to stress and increased disease risk, into a home. Each cat's welfare must be considered carefully, and a solution found which takes into account its needs and what sort of life it would best settle into – whether it is a pet home, in a neutered group of street cats with caretakers, or as a cat supported by someone living in a shed or shelter in a garden or on a farm, etc. We can do these things well if we think it through, explain the cats' needs to others and keep confinement, which is very stressful to cats, to a minimum. Being overwhelmed by the number of cats, the need for funds or the unrelenting work, or doing it badly, even with the best of intentions, can lead to abuse, albeit accidentally.

Can cats give us emotional support?

By domesticating animals, people have developed a role in supporting them. Companion animals can also be a source of support for their owners in that they improve their everyday lives and may help the owner to cope with stressful situations. Cats have been reported to be a source of emotional support for some owners, even if it is just listening without judgement. They are also thought to be able to improve general mood by alleviating their owners' negative attitudes and may make them feel less depressed or lonely, providing companionship and something to care for. Cat ownership is also associated with reduced risk of heart attacks and strokes. But we humans have different needs and personalities and, as we saw earlier, in some research people who rated highly on neuroticism also turned to their cats for emotional support.

Emotional support animals (ESAs) are often now in the news – usually because people claim to be able to take animals anywhere with them because they are their emotional support. We know that pets helped many people cope during COVID lockdown restrictions on movement and interaction. We know that pets may be beneficial to people with mental health conditions, seeming to give calming support and companionship, easing anxiety and providing something to care for and distract from symptoms, as well as giving their owners more positive feelings about themselves. Beyond COVID, studies have suggested that having an animal may boost social interaction and exercise, but I suspect these are dog-based studies where having a dog is seen and reacted to by others outside on walks – cat ownership is more personal. Various research over the years has shown that interactions with companion animals can decrease blood pressure, increase levels of oxytocin (a hormone associated with bonding) and can help with some of the symptoms associated with dementia and Alzheimer's disease.

As I researched emotional assistant animals I came across a picture of a girl with a rabbit on her lap, with the caption 'Emotional support animal comforting a college student'. It did seem that this was all about the needs of the person and not the rabbit, which of course was unaware that it was 'comforting' the student. Had it been transported to the student's college, and did it know this person who was handling it – or was it simply a pet in her home that was comfortable with being handled by her? I hope it was the latter.

In the US emotional support animals have been recognised and can be prescribed by mental health professionals because the presence of an animal is critical to people's ability to function normally and to deal with challenges (in the UK, ESAs do not have legal recognition in the way that assistance dogs do). They are often pets that help to mitigate the impact of their owner's physical or mental health condition through the everyday benefits of human-animal interaction. Animals are not trained but are kept as pets and are different to assistance dogs, which are trained to help people – e.g. guide dogs for the blind, assistance dogs for disabled people. There is a likelihood that emotional support animals are taken everywhere with their owners but that this may not be good for the animal's welfare. Trained assistance animals are developed and trained knowing that they will be out in the world with their owners, and the world of assistance animals is increasing in its awareness of things which may be difficult or stressful for them and how to recognise and minimise this. This could be a good example of use which is not abuse; we can recognise that the dogs are helping people and that on balance it is a good thing.

I think there is an additional pressure on animals, such as cats, that are adaptable (but perhaps not *that* adaptable) to become the focus of supporting someone's mental health. Indeed, even without being named as emotional support animals, in everyday life cats may not enjoy being a constant focus for people and their feelings. Additionally, for most cats, for example, being out and about with someone is likely to be stressful and the cat will have no idea or understanding that it is there for the purpose of supporting the person – if you have followed the discussions in this book about what cats are, their need for control and preferring to initiate rather than react to interactions, you will understand that this is just not a role for a cat.

People crave a strong relationship with their pets and sometimes the need for closeness, control and emotional support in everyday life may become stressful for cats. They are then likely to behave in ways which try to keep people at bay and to regain control (fundamental to a species that has evolved to be self-reliant and may not be socially flexible enough to deal with intense human demands). These are referred to as cat 'problem behaviours', but the categorisation simply illustrates our inability to recognise that cats have different needs to people.

Over the past thirty years or so we have been keen to establish that pets are good for us and our health in order to gain acceptance and provide information to defend them against people who may not appreciate pet ownership or who cite nuisance as a way to control them. We are in no doubt that dogs help people who are blind or disabled and can also assist in medical situations and we know our pets usually make us feel good. However, we may have tried so hard to find reasons to have pets that we have almost forgotten them in the process – now we are citing emotional support animals, perhaps without thought for the animal itself. Why should we have to defend having dogs or cats by saying they are good for our health or wellbeing? Our 'companion animals' are there because we choose to have them as pets. However, while motivations may be genuine, sometimes the outcomes result in negative cat welfare – either purposefully or accidentally. We want to be involved with cats – we have a responsibility to ensure this involvement optimises cat welfare. We need to understand and be honest about what are cats' needs and what are human needs, and align these needs if possible. To take a pragmatic and pro-welfare approach based on understanding cats and the complexities of the human condition!

8.

How to talk to YOUR cat

IT COULD BE argued that when we think of cats as 'pets' we do them a disservice because it allows us to forget where they have come from, what amazing sensory machines they are with their strong hidden drives, and how fascinating it is that they have the ability to fit in with us and enrich our lives. We almost feel they are in this world to be pets, not that they have adapted amazingly to live alongside us and voluntarily stay with us in our homes. If you think about how many people live with cats which are allowed free access in and out of their homes and return every day, how fantastic is it that they choose to be with us, choose to interact with us and fit into our lives so easily.

We are living alongside a creature that is not far from being a wild animal, but which is able to bond with us and has learned to let us know what it wants if only we would listen and take notice. Bearing all this in mind, how do we start to take rather unemotional explanations of body and mind and use them to start to think about how we do talk to our cats and how they talk to us? Here we look at how we can get our minds around our cats and get some ideas about how to interact with them, while the final chapter provides real-life scenarios. In the final chapter, I will recount some of my own cats' behaviours to demonstrate how I have picked up things from them (or what I have managed to notice!) and how different or individual their responses can be.

Why aren't cats obedient?

Some of the things people say if they don't like cats are that they don't do as they are told and are not obedient or loyal like dogs. How do cats react to our attempts to communicate with them? Do cats even know when we are trying to get their attention by calling their names? Recent research from Japan suggests that cats do seem to be able to distinguish that it is their name being spoken rather than other words, and that it doesn't matter if it is their owner or a stranger saying the name. This was based on looking at the pet cats' reactions and at small differences in the cats' responses, such as ear, head or tail movement or vocalisation. Of course, research must take a very controlled look at one behaviour factor at a time and to try and link cause and effect – otherwise it's impossible to guess what a cat is reacting to. Cats do seem to be somewhat sensitive to an owner's moods or emotions and can tell their owner's voice from a stranger's, so they can recognise people's cues in terms of facial or voice cues and body language. Their ability to read people doesn't seem to be as good as that of dogs, but that's not surprising because of the way dogs and people need to live – with others of their species around them. Dogs are able to distinguish different human words and tones of voice from angry to happy, but much of the research relies on dogs following commands or retrieving things, for which they need training in order to undertake the experiment. Not much work has yet been done with cats, so there's a lot still to discover.

In real life, we may use the cat's name but also add noises such as high-pitched sucking in through our lips with a kiss-type noise. Most of us have several different names or nicknames we use, and different people in the house may even use different names. Usually, we use a special tone of voice for cats or a higher-pitched voice that animals recognise as being aimed at them. One of the researchers said: 'Cats are not evolved to respond to human cues, they will communicate with humans when they want. That is the cat.' But I'm not sure I quite agree with this. Our cats certainly did not evolve to pick up human cues, but they do – they know when we are getting ready to go out, when we go to the food cupboard with intent, when we want to catch them to put them in a carrier or to give them medication – even when we are trying to act in a natural way!

How cats respond to our call to action is more interesting.

In many cases cats do respond by coming or paying us attention when called because the call may be linked to something good happening, or they are curious, or they enjoy having attention. They are excellent at picking up our cues – that is how they have learned to train us! We expect dogs to be obedient and come when we ask (or often demand). Why do we want our animals to be obedient? There is of course a need to control dogs because they can be dangerous to others and, in most countries, owners are responsible for their dogs' actions. However, our cats are not a danger and we do not need to be protected against them. Why should we expect them to do as we ask?

Obedience is such a human judgement – why do we do things we are asked to? We act on a request because we want to, because we feel we must, because we would feel guilty if we didn't, because we want to please someone or perhaps we are scared not to. We understand the consequences of not acting if we are asked. There is an inherent need for people (or for dogs) to comply in the way we respond because we have an innate need to be part of a group, and that makes life quite complex. Cooperation and having company are part of our survival because there is safety and support in having a group around you. This requires individuals to understand, to communicate and to be willing to compromise and fit in with other individuals or within a group dynamic, and to have empathy with others. It also sets us up to be lonely or excluded if we are somewhat different to what is acceptable to the group. And because we are tribal, we demand loyalty to particular beliefs or clans.

We value our dogs' loyalty if they show a preference for us, but for cats much of this is not on their radar! Fitting in requires compromise and being led by others. The cat is not being stubborn or defiant or disobedient if it does not do as you ask, and you could argue that it has more free will than a dog, not just in how it is kept but in its thinking processes, because there is no pressure to comply. A cat, having evolved from an ancestor most likely to have lived a solitary lifestyle and neither cooperating to hunt or to defend, does not have these inbuilt needs or the ability to fit in. You could argue that the group-living of some cats (when resources are sufficient) does mean they have to collaborate in some way, and indeed, the care of kittens may be shared, but there is no cooperation in hunting or sharing of food. There must be a reward or at least fewer negatives to living in that way than as a solitary cat, so somehow the cats understand this, but they can revert to being solitary seemingly without worry.

A cat doesn't feel 'guilt' if it does not do what we seem to be asking, but if it does do something with us it is because it wants to, not because it feels it has to. Amazingly, from this base, cats have been clever enough to find ways to tell us what they want and, given the right start in life, they definitely like being with us. However, we have to be the givers rather than the takers in our relationship with cats because we have evolved to be able to do this and they have not. Cats are often classed as 'selfish' and as 'users', but given the tools they have been dealt, they are doing much better at co-living with us than is the case in many people-people relationships and people-dog relationships!

Bearing all this in mind, how can we encourage the relationship between us and our cats (probably for our needs rather than theirs) and convince them to want to be with us?

Let cats take the lead

It seems that it's better to let our cats tell us what they want rather than trying to force ourselves upon them. We can then react and develop a relationship where the cat feels comfortable and is not feeling defensive because we are too demanding – remembering that cats can't see that we are aiming for a closer future with them; they don't understand intention, just what is happening to them. Someone once said that cats are pessimists – they think every situation is trying to kill them! To be fair, to a small animal that can be prey as well as predator, this makes sense. React quickly and get yourself into a position you feel is safer and you can pat yourself on the back, reassured that you have indeed saved your life and will do the same if the situation arises again.

If you have a cat that loves to interact with you, reacts positively to you making contact and even demands more attention, you may wonder what this is all about. However, not everyone has a cat like this and many cats are more anxious, or just perhaps more neutral about the relationship, and are happy to live alongside us without taking part in too much of the cuddling or kissing that we humans love, both physically and as a sign that we love our cats.

Encouraging interaction

Communication is, of course, a two-way process. If you don't listen and watch, you don't know how to respond, and if you don't show willing to communicate, then others may not respond to you. How do we communicate with our cats in response to their communication with us? Do we realise then they are trying to tell us something? It may seem that we should know all of this stuff and that 'researchers' should be able to tell us a lot more. However, the more you understand the complexities of undertaking such research, trying to isolate cause and effect and interpret what you see, the more you realise that all these steps in knowledge and understanding that I've explored in this book are baby steps. This should tell us that we need to watch, listen and observe much more closely ourselves (but without staring or focusing too much on the cat as they don't particularly like that!) and notice what happens when we do things. Think 'spy' or detective – be as subtle as you can in collecting your information and try not to influence the outcome.

It may feel like a bit of an effort to learn some stuff about your cat, but it will make you look differently at it, appreciate it more and interact better. Then see what happens if you respond in different ways. Of course, we may like the way we already view our cats because it allows us to interpret their behaviour in a way that suits us, as something which allows us to love them in the way that we want and to ignore signals from our cats that they may not be wholly comfortable – that's the complexity of the human condition!

The cat's face is not one that gives a lot away. It has very mobile ears, eyes which can also give us clues, and whiskers that are more mobile than we first think. Hopefully most of us don't see our cats in such extreme situations that they are hissing or spitting, so it can be difficult to 'read' them on an ordinary day-to-day basis when the centre of the face does not move a great deal. But in a long-term relationship we have time to observe and recognise the subtleties and react in a way that allows cats to tell us more.

Remember, the diversity of behaviours seen in pet cats in response to physical interaction with people can also be viewed as a spectrum in the lovely diagram produced by International Cat Care in Chapter 4 (page 108). Owners will recognise many of these behaviours in their own cats – I especially love the 'tolerate' illustration, which makes me smile every time I look at it. I recognise it in my cats, usually when it is me who wants to stroke rather than when the cats put themselves forward for it! All of these behaviours can be shown by one cat at different times and in different situations. Some cats will only show some of the reactions in their relationships with people because they are not equipped with the right early experiences to reach 'enjoy' and are more likely to be at the 'avoid' or 'protective behaviour' end. Cats are also likely to develop the ways they communicate with their owners over the years they live together.

Sometimes just how cats sit or lie can also give clues as to how they are feeling. If the front paws are tucked neatly under the cat, bent towards each other or bent back with the pads facing upwards under the chest, they are likely to be feeling relaxed and calm. If, however, the paws are drawn in under the body but the pads are firmly on the ground, it is likely that the cat is feeling anxious, with the paws placed so as to be ready to move if it has to. An example would be when a strange dog is visiting in the house and the cat is sitting up high, watching, feeling vigilant, not relaxed, ready to run away if necessary.

Responding to our cats' sounds

The title of this book is *How to Talk to Your Cat* and, as humans, we focus a great deal on oral communication – we are not brilliant at body language even when it comes to our own species!

The stirring purr
What is more 'cat' than a purr? Being fairly selfish creatures who believe the world revolves around us, we often assume

that our cats are purring because of something we are doing, a reaction to us. If you remember, we know that mother cats and kittens purr when together to encourage feeding and to reassure each other that all is OK. Of course, purring continues into adulthood and adult cats will often purr when they are close or touching each other, such as when one grooms another. They may also purr when they play with a toy or when they're eating, and may indeed purr when they are on their own. However, the purring is usually done in company, and it can be an invitation to care for them, asking to be fed or stroked, or an indication of enjoying interaction with another animal or a human. Cats can also use that purr to encourage us to do things or to continue doing things.

Cats can emit several different types of purr, and one is much more 'encouraging' than the others. They add in a sort of cry or trill to the purr, which is much more energetic than normal, with the intention of stirring us into action. It is thought to move us like the sound of a baby. A study found that when cats use purring to get us to feed them, the noise is 'more urgent and less pleasant' than when they are purring while sitting calmly on our laps – that 'normal' purr is rather more restful and needs less effort. We are actually being manipulated (in the nicest possible way), but it just makes me smile to think of the clever communication of our cats and how they train us when we think we are in charge! They are better students of our behaviour than we are of theirs.

See if you can tell the difference between a demanding

purr for food and a gentler and more contented purring on your lap. Once you recognise this sound you will know that the cat is trying to encourage you to do something – stroke it or feed it or open the door for it – and you will listen for it more carefully each time the cat purrs. And the more you respond to the message, the more likely the cat is to use that method again.

Speaking directly to us

We know that cats do not often use oral communication between each other except in situations of reproduction/ defence or when raising kittens. They do make some sounds, together with small chirps or trills, to other cats they like, but not in the way that they talk to us. Perhaps the miaow is used when kittens are growing and moving further from the nest, curiosity driving exploration but at the same time feeling nervous about being away from the safety of the nest and their mother. They use a similar sound with us as they use to attract their mother's attention.

So, the cat's miaow is very important to us, bearing in mind that some cats can be very talkative and some barely make a sound. We know that some of the pedigree breeds such as the Siamese are known for their talkativeness, but our moggies vary considerably in how much they talk to us and how they use this to communicate. This may be because they are just quiet cats and do not verbalise their needs, or because they have received no response from owners and given up. There is no doubt that if cats miaow and owners look at them to see what they want, then that reinforces the behaviour. Cats are fast learners and great person watchers, so they probably understand us much more than we do them.

Cats are very individual in the way they use the miaow. The miaow, mew or meow has many variations and is uttered with an open mouth that closes on the 'ow' to make what seems like a distinct word. They use the miaow to ask us to do something for them such as feed them, open a door or to get attention. If we break the miaow sound into syllables – 'me-ah-oo-ow' – it is apparent that the cat can vary the length of each component or emphasise one or more to make different sounds with different meanings or emphasis. If the 'ah' component is not emphasised, the cat sounds very sorry for itself, and even more so if it lengthens the 'ow' part too. Repeating the 'ow' at the end draws out the sound and lengthens the plea. If the cat wants something and you move towards giving it the sound lightens, and purring can also be added to encourage you further!

Sometimes it sounds as if the cat is only using the middle 'ah' or it just comes out as a sort of strangled whimper, or nothing at all! There is a book called *The Silent Miaow* published in 1964 by author and cat lover Paul Gallico. The subtitle of the book is *A manual for kittens, strays and homeless cats*; it claims to be a cat's-eye view of homo sapiens and to have been written in the original feline, and translated by Paul. It's a book of instruction for cats, to tell them how to train their families. It's true, cats do sometimes emit a silent miaow, going through the motions but with no sound coming out, and it is indeed very appealing! The miaow is also used to let us know when cats don't like something – like being picked up or moved off our laps, when presumably it is a bit of a complaint or a cry.

Can we communicate by looking or pointing?

Quite a bit of research has been carried out with dogs where people point at an object and the dog follows their gaze to the object. Some work has to be done with cats to show that they do respond to a person pointing at food to help them to find it. Additionally, if the dog sees the food but cannot get to it then it is likely to look back to the person, and step up its eye contact and perhaps add other behaviours to communicate that it needs help to get to the food. Cats do seem to look at people's faces when looking at a potentially frightening thing, and appear to adapt their behaviour to some extent depending on the facial expression of their owner (positive or negative), but a similar step up in communication and help has not been seen in cats in the research that was undertaken.

However, I am sure owners have noticed that cats do step up communication if there is nothing in the feeding bowl to try and get the message across, but this may not happen in an experimental situation. At home cats do seem to know if you are taking notice of them if they are asking for food and so maybe they do look to their owners and up the communication if they are not able to access it. This perhaps shows that this is very situational and individual behaviour.

Blinking can be an interactive tool

Research has shown that if you narrow your eyes and blink slowly at your cat it encourages the cat to approach you and be more open to being with you. Perhaps it's like smiling at cats in their language, and it can't be misinterpreted as a stare (which can be a threatening thing to a cat). When cats are relaxed, they often sit with their eyes partly shut and blink slowly as we look at them. It may be the cat's way of telling us they are not a threat and there is trust between us. Researchers found that cats were more likely to slow-blink at people if they had slow-blinked them (as we explored in Chapter 6). They also looked at whether cats would react to strangers in the same way as their owners and got people to also extend a hand towards the cat. They found that cats were indeed more likely to blink back and were more likely to approach the person's hand after they had blinked. We think that it is a way of them accepting that we are not threatening; an additional tip is to look slightly past the cat rather than directly at it so it can't be taken as a threatening stare.

When you next meet a cat or interact with yours, try slow-blinking and see what happens. We humans like to smile, and when we do this we narrow our eyes, so perhaps smiling brings a signal that we have been unaware of giving our cats. Who knew that a blink was so powerful?

But don't interpret every blink as good – rapid blinking or tight closing of the eyes may be signs of fear or feeling threatened. Cats that blink their eyes rapidly or scrunch up their eyes may be feeling frightened or threatened and in this case it is good to back away so they can relax.

Using play to bond

We think of cats as playful, and play is a great way to develop the bond between us and our cats. We believe play is necessary for development but is also a pleasurable part of domestic life. How much you play with your cat will depend on how much time you want to spend with it and you and your cat's desire to play. Playing with kittens is irresistible, but pretty well all cats, if provided with the right opportunity, will play and benefit from the opportunity to do so, no matter what age. Kittens play more than adult cats, but even older cats may enjoy the stimulation and gentle exercise of a game that is adapted to suit their health, energy and mobility.

When cats play with toys it is similar to hunting and the things that encourage them to hunt are the same things that make them like certain toys – toys which are similar to prey in their texture, size and movement seem to appeal. Apparently, kittens (which love to play with everything to start with) lose interest in objects that do not move at around five months of age. So, owner input to move toys is important, and cats also seem to prefer it if a toy is moved by a person rather than remotely. Cats also like new toys and get bored with ones they've played with before – probably sooner than owners get bored with them!

Does play make cats better hunters? Apparently not: what improves hunting is hunting prey. When we think about play, we imagine athletic chasing and jumping and charging around after a fast-moving toy. However, when cats hunt it is often a slow, steady watching and creeping procedure with only a quick dash at the end – the exercise is in the cat's mind as well as its body. So, play may be one of the cat-specific behaviours they need to express. We can play better with them if we work out what they want to do and encourage them with the right toys, in situations where they can relax and play with us.

You probably have an idea whether and how much your cat likes to play because it makes up games itself, playing with objects, making play/hunting movements with wide eyes and dilated pupils. A cat that is keen to play will play with just about anything you introduce or it may find things itself, even destroying them in the excitement, and will be happy to keep playing. Other cats only want to play at certain times, will not be too bothered about some toys, may need quite a lot of encouragement to play and get bored quickly. They probably don't pick up and play with toys without you.

Toys can include texture, smell, noise and shape to encourage play, but of course the most motivating way of encouraging play is movement (remember in Chapter 1 we looked at the cat's ability to focus on movement and find it hard to ignore). Even simple toys, such as a ping-pong ball, can provide hours of fun and exercise, as can bits of screwed up paper. Cats will also explore bags – and they love a box! Many cats play fetch with their owners, retrieving toys and bringing them back to be thrown again – something we associate with dogs but which seems to be quite common in cats as well. It's a great way to bond with your cat if it is that way inclined. If your cat doesn't go outside where it can climb walls and trees, stalk prey, chase leaves and possibly meet other feline company, then you have a much greater responsibility to give it an outlet for hunting/play behaviour so that it doesn't get bored or frustrated.

Cats like new things and so if your cat has a favourite toy, don't have it out all the time; use it and then hide it away for a while before bringing it out again; this will maintain its novelty. Predatory behaviour usually occurs in short frequent bursts of activity, so perhaps mimicking this makes it more real, too.

Getting play right from kittenhood is important. If play has been rough when kittens were small and owners used hands in grabbing games, kittens will learn to bite and grip strongly with their claws, which is fine when they're small but very painful once they get larger. Therefore, it's important to teach kittens to play gently and to use toys that separate them from our hands, such as those attached to a rod or string.

Lots of toys contain a stimulant for cats called catnip (*Nepeta cataria*). A plant that is part of the mint family, it contains an oil called nepetalactone that causes a catnip-reaction in about two-thirds of cats. It is used in many cat toys – or can be bought loose. It seems to lose its potency fairly quickly, so keeping it in a closed plastic bag may help to retain some of its powers. Whether it acts like a drug such as marijuana or triggers a response similar to a female cat in season or gives a feeling of pleasure is not really understood. Typical behaviours on smelling it include the cat sniffing or licking the catnip, rolling on it and rubbing its head and cheeks on it. They cat may elicit the flehmen response, discussed in Chapter 1, which allows the cat to concentrate on the scent and to smell/taste the herb. The effect may last up to fifteen minutes and then the cat seems to get over it, moves away and may not respond again for several hours. Kittens don't usually respond until they are three or even six months old. It doesn't seem to cause any harm.

If there is more than one cat in the household play can get quite excitable and cats can become more reactive, especially if catnip is added to the equation. Know your own cats and, if necessary, play with one away from the others. Even for cats that enjoy playing together, the excitement can escalate and things can get a little too enthusiastic and tip over into aggression, so trying to keep things calm is always good! I hadn't really thought about this until one of our experts at International Cat Care explained that if cats play around objects like stools or chairs or other obstacles that give them something to move around and hide behind or climb up on, this breaks any staring that can happen and allows things to calm down; the play does not overspill into aggression.

Where to touch your cat

As we keep reminding ourselves, our cats are not truly domesticated and have not totally left their solitary wildcat ancestors far behind. Like them, our pet cats still have the instincts to make considerable effort to communicate indirectly with other cats via visual and chemical messages just to avoid having to see each other. So we know that domestic cats didn't inherit many social skills from their relatives and the skills they have need to be adapted to be used with us humans. We, on the other hand, need to have other people around and like to touch to show love and affection. Therefore, this may not be a relationship made in heaven and we, as creatures that can be empathetic and control our innate behaviour, need to be aware and adapt our behaviour rather than expecting cats to alter theirs.

We know cats are all different and respond to how and where they are touched in different ways. Interestingly, research has reported that even though some cats may react to unwanted attention from us by using behaviours that push us away, a tolerant cat may actually be feeling more stress – because being tolerant is quite a hard thing to do if they dislike being stroked! Holding back on touching can be difficult for people – cuddling and holding are loving and reassuring actions. Most friendly cats enjoy being touched in the areas where their facial glands are located, which includes the base of their ears, under their chin and around the cheeks. These places are usually preferred over areas such as their tummy, back and base of the tail.

The answer is to go slowly and gently and watch for how your cat is reacting – is it nudging you for more contact, purring and relaxing, or is it looking for a way to escape? Signs of dislike include turning their heads away, dipping, moving, or shifting to avoid contact. Even just sitting still with no reaction such as purring or rubbing you means it may be trying to communicate that the action is not being enjoyed. Serious dislike may show as exaggerated blinking, shaking the head or body, or licking the nose. The skin may ripple or twitch along the back as you stroke and short rapid bouts of grooming may also be used to distract and to reassure themselves. Behaviours may escalate to a swishing or thrashing tail, flattened ears or ears rotating backwards and the cat may turn its head sharply to look at you or your hand; it may try to bite or swipe and push your hand away with a paw in order to repel the attention. Some cats do like having their tummies tickled but this is not the 'usual', so go gently and see what your cat likes or dislikes – and don't push your luck!

What about the cat that doesn't want close contact?

People live with cats in a wide range of situations, from feeding street cats to keeping a single cat totally indoors. There is no best way to enjoy cats and many people get great satisfaction in taking the most un-people focused version of our domestic cat and caring for it at arm's length. Many such cats have not had positive early experiences and so don't want to be 'pets', but they may become tolerant and even seem friendly, sitting and waiting until their feeders come, and will greet them. Some may even tolerate a stroke or pat while they're eating, but most keep at a safe distance, ready to dash away to safety. While others may not consider this a close relationship, it is a nurturing one and can be very satisfactory for both sides.

What if you live with an inbetweener, a cat that needs some human help but can't cope with being too close to people – what can you do? Earlier in the book we identified cats that have not really had the start in life that would allow them to relax with people but which still need support and a more arm's-length relationship. Many inbetweeners would do well living a free-roaming lifestyle – unrestricted access outside, where they have food and shelter and a person to care for them from a distance. These inbetweeners require food and water, and shelter that is dry and draught-free. They need somewhere to toilet or they could use the outdoors. They will usually stay away from people but, if they don't feel under pressure, some will start to feel more comfortable and come forward more. First of all, accept that this is how your cat behaves, back off and don't pressurise.

Dealing with any health problems can be difficult with such cats because it is very difficult to get the cat into a carrier if you cannot get close to it. Initiating contact with an inbetweener can reinforce the need for it to hide or even behave aggressively. Making the carrier part of the cat's hiding place or bed means it is in place in case you need to capture the cat. Talk to your vet about the cat and your difficulties and find ways together to do what is best for it. Always let the vet know if the cat is an inbetweener and do not assume that vets have magic ways to handle cats. If they know the cat will be highly stressed, vets have ways of gentle management that do not involve directly handling the cat that they can use or they may sedate the cat if necessary. If medication is required, discuss what can be put into the cat's food because it will be impossible to handle the cat to give it a tablet. You and your vet may need to consider pragmatically what treatments can be carried out. Minor conditions may be treated by putting medication in food but if the cat requires confinement and prolonged treatment, it will be severely distressed under those circumstances and then euthanasia must be considered to prevent or alleviate suffering.

Other inbetweeners need to be near people but don't want to be the focus of attention the way a 'normal' pet cat would tolerate or enjoy. Such inbetweeners may come closer or make contact if the cat is in control and is not pressurised in any way. Allowing the cat to initiate contact can result in more interaction, as the cat is able to be in control.

The kind of interactive style to use with these cats is to not approach or focus on the cat and only give attention if the cat approaches; to have a routine whereby the cat knows when things are going to happen; or to provide dry food on an ad lib basis, so it is always available and does not cause stress.

Extra care for indoor cats

The totally indoor cat, which centres its life and waking hours around the presence of its owners, needs careful consideration. Trends in keeping cats are changing all over the world. In the US, it is almost seen as negligent to let your cats go outside and in some places cats may need to be kept safe from outdoor risk. In many countries people live in high-rise flats in cities with no outside area for their pets. But is there a cost to the cat's mental wellbeing?

The interests and activities of indoor cats have no external component and they can become very focused on their owners. This is the kind of relationship some owners want, but it's one that puts a great deal more responsibility on them in terms of keeping the cat amused, active and not frustrated, especially when they go out to work or even just generally have to leave the home. At the other extreme, cats are still persecuted and cruelly treated in many countries – our own included. Are we worrying too much about a few very well-loved and pampered pets? Perhaps the only thing their owners are guilty of is loving them too much?

While it's true there are still millions of cats that are starving or suffering from disease or injury, that doesn't mean we shouldn't look carefully at how we keep our own cats, at how we can use our knowledge for the better and look at what cats want as well as what we need. Most cat owners fall between these two extremes: they have one or two pet cats that may have access outside some or all of the time and the cats fit their lives in and around that of the family or person in a way that is mutually convenient to both.

It's good for indoor-only cats to keep experiencing new things so that their home does not become too predictable and they get out of the habit of being curious and confident enough to try new things and begin to find them stressful. New things that may smell different or stimulate the cat's curiosity can include the beloved cardboard box that cats always seem to appreciate, paper bags which can be hidden in, different objects and textures, things hidden within other things and placed around the house for the cat to explore.

What type of cat do you have?

Once you have read this book perhaps it is time to think about the type of cat you have – is it confident and relaxed with you and your family? How does it feel when strangers come to visit? Where does it spend much of its day? Does it hide away, or does it run forward when something new and exciting is happening? How does it react to you? Does it duck under your hand, or does it jump up to meet your hand as you reach down to pat it? Of course, cats may do all these things at different times, but what is your feeling for how often they occur and under what circumstances? Does the cat behave differently when other cats are around? There are no good and bad or right or wrong answers; just thinking about them and answering honestly is a way to understand your particular cat.

Does having a number of cats affect the relationship?

Many of us have more than one cat and this can have an effect on our relationship with them too and we should be aware how this adds to the mix. If cats get on with other cats in the house and they are happy to sit together and groom and maybe play together, then the relationship can be relaxed and not impact on either of the cats or upon your relationship with them. Indeed, they may both try to sit on your lap at the same time! The cats may not be hostile towards each other but manage to tolerate and mostly ignore each other. However, if there are tensions, which is probably more often the case than not in households with more than one cat (and the likelihood increases the more you have), this can impact on many things.

More often than not we have decided to add another cat to

the household because we would like one (and there is nothing wrong with that), but we do not know what the original cat wants, and then we choose the new cat to add into the mix. Consider how you would feel if someone were asked to join the family and you had no choice over whether it happened and who the person was. You would then be expected to get on well! And remember that as humans we might have a more inherent need to make it work, to reduce stresses within the household and to keep everyone happy than a cat, which feels no such pressure to compromise.

As a general rule, the more cats there are, the more likely that some will not get on with each other and tensions will overspill into others as well. It can be the most major reason for unhappiness for cats, and for people who just want them to get on and be friends! Once you recognise that cats are not getting along, you can do some things to give them a chance to get away from each other by providing hiding places and high areas where they can feel more relaxed because they can see what is going on. This may be quite a complex situation and if cats are soiling in the house or fighting you may want to think about seeking the advice of a cat behaviour specialist or even consider if there is a better place for one of the cats to live, where it is not going to be stressed.

What sort of owner are you?

Are you someone who needs to cuddle and kiss and wants to

show off your cat to visitors? Are you the kind of person who will let your cat tell you things by watching it carefully? Are you determined your cat will love you and will continue to try to convince it to do so no matter what? Unfortunately, having a previous cat that behaved in a different way may not have been down to you but to the cat itself, and no amount of persuasion will change your current cat's behaviour. It can be an unsatisfactory relationship if close interaction is vital to you and your cat seems to have the opposite feelings.

Many of these questions probably ought to be asked before you take on a cat so that you can try to find a confident kitten that is happy to be handled and meet new people – The Kitten Checklist, available on the International Cat Care website, can help prospective new owners to ask the right questions when choosing a kitten. If you are taking on a cat from a homing or adoption centre, then you can try and ascertain how it interacts with you and you can talk to the staff who look after the cats and get some idea of its personality. But remember that a stressed cat may not be at its best in terms of relaxing and letting its true personality shine through – if the cat is in crowded conditions or in a centre which has barking dogs and there is nowhere for it to hide in its cage, then it may be too distracted and anxious to interact at all. Giving cats the best opportunity to interact in a quiet and unthreatening environment may allow you to see what they are really like. However, at the end of the day, we know that we take on cats for lots of reasons, not all of which are under our control!

Learn from our cats

We have to be detectives with our cats and gather small clues as to what they like and dislike, to watch and listen to what they are trying to tell us. We need to respond to our cats' leads rather than taking the lead ourselves, but at the same time be willing to let the cat know that we are open to being friendly if they wish. If we respond and then take note when 'enough is enough' in terms of stroking or cuddling, they are likely to come back to us if we do not overwhelm them.

When we do, we can start to react to them in a way which is sensitive to the signals they are giving and supports what they are asking for. The more we do this, the more we build up trust and a positive relationship that can mean our cats turn to us more for attention, confident that we are not going to do something which makes them anxious.

Cats teach us to be positive with our interactions, to think creatively, to be sensitive to their cues and to go gently in our interactions with them. Punishments will not work to make them comply but are likely to cause them to be fearful of us and to withdraw from interaction. Because they are not built for group compromise, they will not understand or react to the negative behaviours which both people and dogs may put up with in order to stay as part of the group. We get away with a lot with our dogs because of their need to stay with us; our behaviour needs to be much better with cats. Perhaps then our behaviour with our own kind might improve too!

9.

Meet my cats – how do we talk to one another?

A LOT OF this book has been about research and what we understand about cats in general but, as we have seen, cats are individuals. I thought a way of illustrating this and looking at how reality shapes up against some of the theories would be to look at my three cats, which are all very different and have developed their own ways of training us! This is not a science, but gives examples of individuality and ways of viewing the whole cat and what it is doing in the context of how old, healthy, confident, nervous each cat is and the way it reacts to certain situations. It's like taking little pieces of a jigsaw and putting them together to give a fuller picture.

Living with us

As I have been writing about cats in general, I have been considering how my own three cats behave in light of the research and the insight we are gaining about cats. I've been thinking about the differences between them and their individuality, considering they have a common stable home and common experiences. I have had them all since kittenhood (albeit with different early starts in life).

Two are now aged nine and one is ten; their characters have stayed very consistent over that time, although their behaviours with us have evolved and their characters become clearer as they have aged. When they were kittens, I had children aged around sixteen and ten at home, both of whom were gentle, sensitive and patient with them. My partner, not initially a cat lover, is great with them and is as interested in figuring them out as I am. He has come to appreciate them as individuals, built his own relationship with them, and they often prefer him to me! All three cats would fit onto the 'pet' cat end of the lifestyle spectrum that we observed in Chapter 2, although one of them could easily have lived as a street or barn cat, had we not got him at an early age.

We are lucky enough to live in the middle of the countryside away from a main road and with a large garden. The cats have free access to the outside via a cat flap. We are near a railway line, but they stay away from it and prefer to explore the large quarry area behind the house instead. Our near neighbours do not have cats, although there are a couple in houses about 300 metres away, so there is no close competition, which is lovely for them. There do not seem to be many other stray cats in the area – occasionally we see one in the garden that seems to be a farm tom cat on the prowl, but not often, although I suspect the cats will be much more aware of strange cats than we humans are. They have lots of space indoors and places to get away from each other if they wish. They all go outside as much as they wish and unfortunately are pretty good at catching rodents, which they bring in with regularity, especially during the spring and summer.

Meet the cats themselves

Meet **Chilli**, our oldest cat, who is about ten years old. He was found when he was a very small kitten (we think that he was about six weeks old), running around in our local country pub car park, perhaps coming from a neighbouring farm, where kittens had been seen before. My daughter managed to catch him and save him from someone who was trying to run him over, and brought him home. The immediate response to our ministrations was a large and vigorous purr from this tiny ginger bundle while he sat looking at us in my daughter's cupped hands. Of course, we could not resist! I have never had a ginger cat before (previous cats had been mostly black and white or Siamese) and I was curious about their reputation for being friendly. Ginger cats seem to have universal appeal, perhaps for the attractive coat colour and pattern but also because of their reputation for having a warm personality.

Does being ginger affect personality? In Chapter 3 we looked at individuality and discussed coat colour and genetics and, although it is not clear if there is a connection, there are some ideas about how it might have an effect. I also found some research into colour and behaviour in horses which compared bay and chestnut horses (chestnut being the equivalent red colour in horses). They found that chestnut horses did not display any difference in behaviours such as rearing, biting, kicking or bucking when compared to bay horses. However, they did show more 'bold' behaviours (remember we defined bold as showing a willingness to take risks, and being confident and courageous) and were more likely to go up to unfamiliar things in their environment, and so were shown to interact differently to their surroundings compared to, say, bay horses. Studies of people with red hair have found that redheads feel pain differently and have different body reactions when compared to blond or dark-haired people, and may require different doses of drugs such as anaesthetics. So there may be something in the ginger cat folklore and perhaps ginger cats are bolder and therefore more willing to take on new experiences and put themselves in a vulnerable position that may allow them to find that humans are quite nice!

If he had been born on a farm, which was my instinct, Chilli probably hadn't had much interaction with humans, and I would have expected at least some timidity in our home and with all of us crowded around him. But he was not fearful and simply fell happily into being in the house with us. He was very young and probably still within that period when kittens are more open to learning about other species and coping with new things. Did he inherit genetic friendliness from his mother or father? Did his bold and confident personality (perhaps aided by genetics linked to his colour) overcome fear despite not having a great deal of interaction earlier in life? It could be that he had escaped from a home and had been handled before, but we didn't hear of anyone who'd lost a kitten so will never know, and I am always grateful for having him with us.

Compared to the other two cats he is street-wise and he would also be the one who might scratch or try to bite you (a gentle nip) if you pushed him too far (which of course we try not to). He is more than likely farm cat stock, with a thick coat that could never be described as smooth or shiny and gets thicker in the winter even though he luxuriates in the indoor heat as much as his housemates. He seems to have a lower threshold for reaction and higher instinct for self-preservation, should he need it. He is also the one, however, who can be turned upside down in my children's arms and will relax so much when his head is massaged that it hangs down as he purrs himself silly, just as he did as a tiny kitten. Neither of the other cats would put up with that! So, he seems to come with an inner confidence, but takes no nonsense if he doesn't like something.

He is a watcher and seems to have the wisdom to watch occasional strange cats outside without actually going out to interact with them, although when he chased a fox (which was exactly the same colour as him) out of the garden we held our breath for his safety!

At the time of finding him we had a lovely elderly black and white cat called Diamond, who was at the end of a long battle with hyperthyroidism. I had not planned to get another cat until she was no longer with us. At the time I suspect she probably didn't appreciate his presence in the background, but we decided to keep him because we knew she was quite unwell, and he was exactly the cat we would have sought when she passed on. We kept him from annoying her, and indeed she was pretty impatient with him if he came near, even though he never did more than look and try to be friendly. He didn't seem to be offended by her rejection and just joined in with the people instead. Sadly, Diamond did leave us soon after that, but having Chilli definitely helped us with the loss.

Chilli at ten years old has developed in line with the confident and curious small kitten we found. He has continued being a big personality and loves joining in with what we're doing, earning himself the nickname of Mr DIY for trying to get too close to all home decoration/renovation activities. This big personality seems to just about be contained, as you will see when we discuss his body language; and he is usually on the move and can't sit still – we say he is looking for mischief, but he just seems to be constantly curious (perhaps even nosy) and is interested in everything that is going on. He was one year old when we brought home two new kittens and he accepted them with curiosity but calmly and gently.

When he was seven years old Chilli developed a urinary tract blockage that was life-threatening. In this condition the blockage prevents urination and the urine builds up in the bladder. Not only is it a problem because the bladder gets fuller and fuller, but the toxins in the urine affect the body and can damage the kidney and heart and cause the cat to feel very unwell. Unless the blockage is relieved quickly this can be a fatal problem.

My friend Vicky Halls had been staying with me and we came home and found Chilli lying on the settee. We took one look at his face, which showed he was in pain – and he was lying in a tense and strange position – and we knew something was definitely not right. Vicky had been a veterinary nurse as well as a behaviour expert, so between us we gently got him into a carrier and took him to an emergency vet as it was late in the evening. We guessed he had a blockage but obviously did not know exactly what the problem was. The vet managed to position a catheter and allow the urine to escape and the next day we then went back to our own vet to try and stabilise him and ensure the blockage was gone. However, our vet was unable to keep his urinary tract free when the catheter was removed and referred us to Bristol Veterinary School for him to have surgery with a specialist. He had to stay there for a while to make sure the surgery had worked and my daughter and I went to visit him there – he was very much enjoyed by the veterinary staff who cared for him because he purred a lot and was quite a character.

Chilli made a great recovery. He has had no other problems in the years since then and I am grateful to all the vets and nurses who cared for him so well. I was also very pleased that he was insured!

Now meet **Mello**, one of the two kittens we got when Chilli was about a year old. She is a small and beautiful brown tabby girl and came with her brother, Oreo, when they were eight weeks old. They were born on a local farm from a white mother cat who probably had some pedigree breeding somewhere in her background. The father was likely to have been a farm cat and the four stunning kittens in the litter were ginger, ginger and white, tabby, and pure white – a most beautiful group of kittens. Both she and her brother have short and shiny coats. They had been well handled as kittens and having been born in the owners' bed, looked upon people as a normal part of life for them from the very beginning. Both were confident with people, curious and very happy to sit on everyone's laps and sleep on their shoulders when they joined the family. I didn't need to ask Mello's sex when we went to see the kittens – her little face was so petite it was obvious she was a female cat. Mello still loves people and sometimes even moves into that 'needy' category – but more on this later.

Oreo is Mello's brother, not that you could tell from looking at him because he is pure white with green eyes and is much bigger and sturdier than her. Mr White, as Oreo is more often called, is what someone once called him 'nice but dim'! He is not hugely interactive unless he wants to be, but he is calm, strong and quiet and ignores most of what is going on around him unless he is directly involved. He is the least interactive of the three when he is doing his own thing and is happiest out hunting and then sleeping on the bed. However, that's changing as he gets older and he is getting more demanding about climbing on a lap and being with us.

Mr White is the king of sitting down! Even as a kitten he would sit or lie down almost immediately he arrived at where he wanted to be, even if that was in the middle of the floor. He loves my daughter and has a close relationship with her that extends to being cuddled and kissed, which he does not accept from other people. He sleeps on her bed when she is here and sat on her lap when she was a student and working at her desk at home during lockdown.

Mr White also had a health crisis only several months after Chilli had his. He went downhill very fast and my vet called to tell me that, unbelievably, it was a very similar problem to that suffered by Chilli and it was best to refer him to the university hospital as well. He had to have a blood transfusion and surgery to insert a plastic contraption called a 'sub' (a subcutaneous ureteral bypass) to overcome a blockage between his kidney and bladder before he could come home. It's quite a new intervention and vets are still learning about it and ways to improve it. He had regular visits to the referral veterinary hospital in Bristol to ensure it was working and to 'flush' the system but, miraculously, had it removed after two years because the stones blocking the system had disappeared. I asked my feline veterinary specialist friends if there was any connection between both cats suffering this problem so close to each other, but they felt it was just a strange and unlikely coincidence.

Mr White is now back to normal and he remains a calm and happy fellow. The care of both cats was undertaken at Cat-Friendly Clinics, where they were handled with respect and gentleness by people aware of the stress cats go through in such situations. They both seem to have come through the experience very well and have stayed well ever since.

Relationship between the cats

I watch all the cats with interest, not just their individual behaviour but how they interact. Although Mello and Oreo are brother and sister, have had almost identical and good experiences as kittens and are neutered, they are very different individuals. They are not the best of friends and Mello prefers Chilli to her brother.

The best relationship is between Chilli and Oreo, who are not related but regularly play, groom each other and curl up together. We know that when cats are happy to be together they often groom or rub against each other, usually around the head and neck. As we saw in Chapter 5, it is usually those more dominant or higher ranking which do the grooming of lower-ranking cats. If you have several cats you may think that one is rather less likely to back down than others or is a bit more assertive about what it wants (and usually gets), or may seem to bully other cats to push its agenda forward – that is probably the one doing the grooming.

Chilli does the grooming, with Oreo often approaching with his head down and presenting himself for grooming, which he seems to love, purring loudly in an encouraging way with that little shrill note added. Chilli usually obliges and grooms, often accompanied by clamping his spread paw and claws onto Mr White's head as if to hold him still. In nature cats only groom each other for a short time and the cat being groomed sits very still. Mr White does indeed sit very still and sometimes the grooming goes well, stops and both fall asleep. However, sometimes Chilli cannot resist nipping the ear he is cleaning and then it all falls apart with a bit of a tussle! However, it never puts Oreo off trying to be groomed and approaching as close as he can get to Chilli. He does not groom Chilli, which fits with the theory of the more assertive cat doing the grooming. Mr White never shows any behaviour which tries to get the better of Chilli, often plays with him and seems very happy to be on the receiving end of a rough tongue!

The boys also play together, and the end result is similar to that following grooming – sometimes it ends gently and sometimes in a tussle. I remember my mum used to say 'calm down or it will end in tears' when we kids played with a bit too much excitement – the same is true of the Chilli/Oreo play. I listened to a talk by cat-behaviour expert Sarah Ellis on play, which illuminated these behaviours. She explained that play tends not to escalate to rougher play (which can also lead to a growl or a hiss because it's getting a bit too rough) if there is something to play around or hide behind, because this helps to break the intensity. Our cats use a tiny little footstool, which is about 20 centimetres high and which is in prime position next to a radiator in the winter. My mum embroidered the top so it is very satisfactory for getting claws into as well! One cat gets on top of the stool (Chilli, of course) and one goes completely underneath (Mr White). Mr White then taps or grabs Chilli by creeping out from underneath, and pops back beneath. It gets fast and furious, but as long as they play around the stool and can get away from each other briefly, it remains excited play. If it progresses to open ground then it can get too rough and after some swearing at each other they walk away, have a groom and carry on with life. It's a great way to watch body language that you don't normally see – turned-back ears, sideways bodies, swishing tails and staring at each other. Even if it ends in a bit of a stand-off, it doesn't seem to affect their warm relationship at all.

Mello is not fond of either of the boys because, on rainy days, they seem to get a bit bored as they don't like to get wet and then get restless and are more likely to chase her to keep themselves amused. They may also try and move her off a favourite spot if they want to sit there. They don't hurt her, but she is constantly aware that this may happen, and I am aware that she would probably much prefer it if they weren't there at all! There is definitely no grooming or curling up together going on between Mello and either of them. They will all sleep on the settee but with space between her and them, and they will choose to eat together or wait together to welcome us if we have been out of the house for a while. However, Mello is more vigilant than I would like her to be. When I picked Mello and Mr White up from the farm where they were born, there were three male kittens (including Oreo) all playing rough and tumble in the dog's bed, while Mello had tucked herself quietly into a corner away from them – not much has changed there then!

When we look at the interaction diagram we discussed in Chapter 4 (page 108) it is interesting to see which behaviours my three cats use. Mello's responses are in the majority 'tolerate', 'enjoy' and 'need'. She does do 'avoid' if someone tries to pick her up or cuddle her too closely, which she is not fond of. I have not seen any of the 'protective behaviours' from her towards people although she does exercise them if the other two cats approach with mischief. Mr White uses 'tolerate' and 'enjoy' (but not 'need') when he does come to you, but he may spend more time independent. Saying that, he has recently got quite demanding about coming onto your lap when he is ready to and as I type this with my computer on my lap while sitting on the settee, he has pushed into the space between the edge of the laptop and me and put his head on my arm while purring, which is making typing rather difficult! He seldom uses 'avoid', probably because he stays out of the way unless he actively wants interaction and is rather stoical and tends towards 'tolerate' rather than 'avoid'. 'Protective behaviours' rarely come into action except when he and Chilli's play or grooming gets a bit excitable, but I have never seen it used against people despite having to put him in carriers to go to the vet quite frequently.

Chilli does lots of 'enjoy' when he wants attention but is also good at 'tolerate' when he doesn't want to be stroked. He will dip under your hand so you miss him – I would class that as 'avoid'. It's funny how accurate these small moves are, which just keep you out of range! He is usually on the way to do something, so is focused on whatever that is and isn't put off the mission. I was discussing this 'dip' with a friend who has a Labrador dog, a breed famous for being ultra-friendly, for seeking attention and for always being hungry and eating anything. 'Surely he never dips?', I asked. She replied that when she is feeding him and about to put his food down, chatting to him and patting him, he does indeed dip to avoid it because he is so focused on the food bowl and getting his dinner. So even the ultimate friendly dog can dip sometimes! Chilli also does 'need' in certain circumstances which are peculiar to him and when he is as determined to interact with the same single-mindedness he shows when he is off on a mission. This may happen when you are working at your desk or getting dressed in the morning, when he will follow you around, poke you with a paw and purr loudly for attention.

What are their relationships and 'tells'?

Look at the body language of your cats in your home – hopefully you will not see the more visible but extremes of behaviour, which may only show if there are heightened tensions between cats or they are fearful of people. Unfortunately, other behaviours may be more subtle and therefore more difficult to interpret. In previous chapters we have looked at the positions and directions of body, head and ears, but each cannot be taken in isolation as we need to look at the whole body and the circumstances of the behaviour before we can really try and figure out what's going on.

People who play poker speak of 'tells' – unconscious actions people do even though they are trying to conceal what they are thinking, perhaps even attempting deception. We often say that cats are masters of disguise, able to hide signs of illness or pain so as not to give any signs of vulnerability, and because they do not need to live with other cats and their communications mostly try and keep them at a distance, they have not developed overt facial signals either. It is not conscious deception, more a lack of signals. If we can notice these more subtle and hidden messages it is as satisfying as noticing a tell from a poker player! Hopefully they are insights into what our cats are thinking, even if they are not purposefully hiding their thoughts from us. Consider how bad we are at figuring out human body language – we do pick up some subconsciously, but many pass us by without us noticing or realising the significance. Perhaps if we learned to watch our cats better, we'd be more sensitive to the body language of our own species!

We have now had our lovely cats for nine and ten years, so we are all very familiar with each other. My kids have grown up and moved out and now only my partner and I are in the house most of the time. Let's look at our cats' particular 'tells'.

Mello is a very sociable, people-loving cat and she very much enjoys being stroked. However, she does not like being picked up or being cuddled because that takes control away from her and she may have to make an escape – however unlikely we may think that is, her instinct tells her that is the case. Yet sometimes she craves attention and there is probably some conflict between these two states. She is slightly 'on her toes' because she doesn't like it when the boys might chase her. She will grumble at them if they come along looking as if they are up to something! She is much more wary of her brother Mello than of Chilli, who she seems to like and will sit quite close to him and approach him to sniff, although they do not rub against each other.

If you watch Mello you will realise she's an 'ears' girl. When she is contemplating things, even when she seems relaxed, her ears are moving in response to every small thing going on. When she is alert, they move and flicker together as well as individually and you can almost see that something is going through her mind – if she hears a noise she is wary of, her ears turn around and back, but soon return. They are mostly forward-facing and alert and they remain that way if you stroke her and as she studies your face to see how you're reacting. They seldom fold back fully and only go to that 'u' shape when she gets more worried or conflicted, returning quickly to forward-facing when she is reassured. She is a fast reactor – if you make even a tiny action to move when she's on your lap, she is up and gone – in our house if Mello is on one of our laps we have to ask each other to pass us things gently because any move may make her jump off!

Chilli, on the other hand, is a 'tail' guy. Textbooks tell us that movement of the tail may signify that the cat is angry or not happy, that it is probably not a good sign and should be a warning to stop what you are doing with the cat. In general, this is pretty good advice to avoid escalating behaviour which might become defensive and it's wise to keep out of the way of a cat that is aroused. However, again, we have to see this in the context of our cats' personalities. I say this because Chilli wags his tail frequently; in fact it moves from side to side in almost everything he does. It varies from sway to wag and sometimes to swish! It's a long thick tail that would drag on the floor if he didn't turn the end up. He's not an angry or aggressive cat at all; he prefers to watch but stay out of trouble. He's very curious and loves being with people, seeing what they're doing.

As with Mello's ears, I consider that Chilli is thinking when his tail is slowly wagging, and it does move most of the time. When he's watching birds in the garden it's swinging and twitching; when he's deciding where to go it is moving – it's his thinking 'tell' and when he hears or sees something, it moves. He's also very tactile with his tail, wrapping it around your legs when rubbing against them. The end curves strongly around you as if it is a hand. Both my partner and I have experienced his 'exploratory' tail – if he is sitting on the desk or on the bed moving his tail slowly and it comes into contact with your hand, he twitches it quickly away as he knows he has touched you, but then he continues to move his tail back and forth, touching your hand every few swings as if he is just checking it is still there and then pulling it away. It's a very controlled movement and exploration with his tail – the other two cats do not use their tails in anything like the same way, holding their tails pretty still in most things they do.

It is rather a magnificent ginger tail, which is a bit paler at the tip, so it's a major part of Chilli's personality. Chilli's ears don't move to the extent that Mello's do. His whiskers, however, also give a lot away. When he's keen to be with and interact with you, they are almost bristling forward in a way that, again, the other two cats' whiskers do not; I have found myself looking at them to see if they use their whiskers in the same way. They do, but not to the same extent or with the same enthusiasm. It's almost as if Chilli is bristling with enthusiasm and energy inside, but in keeping it all contained it leaks out through his tail and whiskers!

Oreo, or Mr White, has far fewer 'tells' and is much less demonstrative. He doesn't have Mello's ears or Chilli's tail tells and moves through life quite quietly and confidently, happy to be friends with Chilli and slightly subordinate to him. If you want to get some ear movement from Oreo, put the hairdryer or vacuum cleaner on – he won't run away but just turn his ears back in that classic 'u' shape, showing his dislike and looking rather confused until it's turned off. He is solid and straightforward and he can get your attention if he wants to, but otherwise just gets on with life.

Getting our attention

Our cats have free access in the house and outside, so the only thing they have to ask for (if it's not already available) is food and attention, or occasionally having to let us know something is not quite right. And, as you can guess, they all have their own individual approaches.

Chilli uses the rubbing approach when he wants food or attention, literally wrapping himself and that wonderful tail around our legs and tripping us up if we try to walk. He does want you to walk towards the food bowls to replenish them because they are empty, but as he is very busy persuading you to take notice, you have to break free, guessing that is what he's after. The confirmation that you are on to the right thing is that he will take his paw and gently try to trip you up as you walk towards the food, and has recently started, at the age of ten, to make rather pathetic little miaow-like noises at the same time, just to make sure! He's not a big talker normally and his sounds are quite quiet and soft, like the little nickers and peeps he makes when trying to scratch open a door if the way is barred to him and he wants to get in.

When you think of how cats use their paws for grooming, grasping food, exploring items and defence, it shouldn't be a surprise to learn that they can adapt their use with us as well. In the UK there is a television advert where cats take over the world because they have developed an opposable thumb and so can grasp with a paw-like hand – the success of the ad probably lies in our realisation that this could be true! Chilli's use of his paw has developed over the years and is now also used when he wants attention in other circumstances, when he has learned to touch or tap us with it, purring heavily and asking for stroking. There is no dipping out of the way then, but instead he pushes forward with his head and shuts his eyes seemingly in an ecstatic way. If you rub the area at the top of his nose and between his eyes he will seem to almost go into a trance.

You can also tell when he comes across something he doesn't like. He shakes both front paws one after the other and walks away from a wet patch or from food he has explored and doesn't like or a smell which is offensive. It's a very strong 'don't like this' message! He also has what we might call 'moments of excitement' – bouts of rubbing and purring furiously that are now almost predictable – they may happen when you are in the bathroom or trying to do a video call. What triggers these or why they happen where they do, I'm not sure. Like other cats he also gets very excited on the stairs and is likely to grab you through the bannisters or grab your hand a little too roughly if you put it through to pat him, so it's always best to leave him alone on those occasions. It is, of course, a position of control to prevent others going up and down, too. If he meets you at the top on your way down the stairs, he always has to run down quickly to beat you to the bottom like a child. Don't we love our cats' little foibles!

Mello, on the other hand, is very noisy and can vary her miaow from a quiet request to a rather long and drawn-out moan with an irritated edge to it if she's feeling really determined and not getting what she wants. It can vary from very pathetic to rather cross. The 'need' illustration (page 108), for example, is very similar to how Mello now tells us she wants some attention. She has gradually developed a repertoire that has resulted in her sitting up on her back legs and reaching up with her paws to attract attention while looking directly at your face and miaowing. Indeed while writing about cats not liking staring I do think about Mello's habit of looking you directly in the face and into your eyes to get your attention and perhaps to see how your face reacts. She obviously doesn't feel this is an aggressive stare from her to us or from us if we look directly back at her. If you put your hand down (which of course you can't resist), she holds the hand with her paws (no claws) and gently pulls it towards her head to stroke her. Some research has found that cats move their bodies to align us with the places they like to be touched, that they prefer being stroked around the head and close their eyes if they like it. Mello certainly pulls with her paws and aligns her head as well so that you stroke under her chin and she can rub against your hand.

Mello is also very interactive with the people living in the house and loves it when others come to visit. It looks like we have taught her a party trick when she sits up and puts her paws up, but she has developed this all on her own, and of course it gets a 100 per cent response as it is so very appealing. She has trained us to respond rather than the other way around and each time we do, we reinforce the behaviour. We, of course, can't resist making it happen by waiting just a beat too long when she is sitting in front of us and miaowing, so that she has to go the extra mile and sit up. Mello also loves to sit on laps and loves a visitor; they will not be discriminated against because they are new and, in fact, she may demand more attention if it's someone fresh! Sometimes it can seem as if we never give her any attention and she has to look to someone new to give her what she desires.

Her 'moments of excitement' occur when there's a jigsaw puzzle on the table that is completed over a period of evenings or days – she knows as soon as you go to the table that you are going to sit down and will be available for attention. Even as you head towards the table, she leaps up onto the puzzle with a little murmur or trill and proceeds to purr like mad, lick the hand trying to do the puzzle (which is very irritating!), sits on the puzzle and, if no stroking ensues, gently pushes pieces off the edge of the table with her paws until attention is paid!

Mr White, as you have probably already guessed, is not a chatterbox or a paw user. However, he is a great greeter and will leap up onto his hind legs if you reach down to pat him when you haven't seen him for a while, bumping your hand with his head – just the once and then that's enough. When we were trying to film this behaviour for a video we were making for International Cat Care, I had to ask the cameraman not to stroke him because when Mr White has done it once it doesn't get repeated, and we needed to catch the one leap on film.

When Mr White wants food he 'shouts' a loud and single meow, looking intently at you and repeating it loudly several times if you don't respond. There is no missing the conversation! He is then silent again but does purr as you put the food down. When he visits us on the bed in the morning he jumps up and does that massive demanding purr with the extra dimension and treads very enthusiastically with his paws until he is stroked a few times, then he lies down and sleeps the morning away. He used the same approach once when, unknown to us, the cat flap got stuck and wouldn't open and he wanted to go out. He almost imitated the famous 'What's wrong, Lassie – has someone fallen down the well?' scene from that old black and white film where the dog barks at its owner until he is followed and then leads him to where something is amiss – in that case someone had indeed fallen down the well. As there was food in the dishes it obviously wasn't that Mr White was shouting about – he walked to the flap miaowing loudly until we realised what needed to be done. Who says cats can't communicate! We were impressed.

I've been watching when our cats scratch – it's thought that cats in multi-cat households may use scratching as a message to other cats and also to their owners, perhaps to get their attention. Chilli is chief scratcher in that he will do it on the stairs (which excite him a lot) and on the bed before he jumps up (I am not sure this would happen if we were not there, so need to try and work that out). We bought leather settees on purpose, knowing his propensity to scratch material, so he has left them alone! However, he loves to get into our summerhouse, which is usually locked but is full of furniture (covered in very scratchable materials) and junk at the moment. As soon as the door's unlocked he appears from nowhere and goes straight to the furniture to scratch as if he's not had access to a nice bit of scratchable material for ages! Mello will scratch on the table leg next to where they are fed while the food is being put down and also loves a bit of scratching in the summerhouse. A lot of this seems to be about exhibition as much as a need to sharpen claws. Interestingly, Mr White doesn't scratch anywhere in the house – perhaps his quietness is the result of a solid and quietly confident behaviour with nothing to prove!

How do we touch our cats?

All of our cats have been handled gently and with respect for their entire lives and we don't try to wind them up, don't attempt to touch tummies, and we take our cues from them. Most cats are defensive about having their tummies touched – in a fight cats may try to injury others by raking their back feet down the stomach of the other cat as it is a very vulnerable part of the body. However, there are some cats that are relaxed enough to let people rub their tummies and will lie with their front legs above their heads and enjoy the interaction. Work this out slowly and gently with your own cat – you will soon get the idea of how it feels about it! Ours are not impressed and such behaviour may end with us having our hands clutched in claws, or cats moving away quickly to avoid the situation.

Cats often groom people and perhaps our stroking is received as grooming by them. They prefer us to groom around the head area and will lean into our hands and sort of direct them to where they want us to rub. They don't seem to like their paws being stroked or even touched and will jump off our laps if we try to interact with their very cute paws. You can hold the paw gently and maybe stroke the top, but more than that is obviously very irritating!

We all want to kiss our cats but often cats are not too keen!

I watched my daughter kissing Mello, who was sitting at the end of the settee with my daughter leaning down over the arm heading towards her – Mello's ears both flattened out of the way even before they were touched and popped up again once the kiss was over. She puts up with it because she loves getting attention, but I don't think it's her favourite thing! My daughter and Mr White have a very solid and simple quiet relationship, and he often sits with her as she works. He now seems to tolerate or even push his head towards her when she kisses him, which I don't think he would do with the rest of us! I think she subconsciously signals that she's going to kiss him and he's usually standing free rather than being cuddled and is prepared. He even seems to enjoy it, likely because he has the choice to get out of there if he wishes! There is obviously a lot of trust and gentle interaction that has been built up over the years and which he is comfortable with.

Chilli may put up with a kiss on occasion but usually ducks out of the way! None of them seek to be picked up and will put up with it a bit, but often with a bit of a miaow moan. Chilli, typically for him, will avoid it, but when he does get picked up and held in a way that he likes, he perches on top of one arm and is steadied by the other hand (so not held tightly) so that he can look around and see better from the higher position. It reminds me of how those little pocket-sized dogs love to be high up to see things from this point of view rather than being at ankle height all the time – many often get quite bossy or aggressive from that position, buoyed up by their owner's presence. Chilli sits quietly and simply looks around. Our previous cat Diamond always hated being picked up and would miaow forlornly – something she did even when a very small kitten, so she was spared this unless we had to pick her up for a reason such as putting her in a carrier.

What about the slow-blink we talked about earlier? We think that it's a way of cats accepting that we are not threatening them by breaking a stare. The slow-blink is subtle, and it's best to look slightly past the cat rather than directly at it so it can't be taken as a threatening stare. However, we aren't really sure. I thought I would try it out on my cats to see what happened. First with Mello, who is a cat who looks you in the face frequently and for long periods of time while miaowing to get your attention. She seems to be studying you for reactions and doesn't mind if you look back – perhaps seeing how you are going to react is more important to her, she doesn't seem to interpret it as a stare or threat. If I sit with her on my lap looking at me and blink slowly and in quite an exaggerated way, after a couple of these she will half-close her eyes, then close them slowly, half-open them, and often then close them fully.

Interestingly, I have tried to make eye contact with both Chilli and Mr White to see how they react, but they don't engage at all in looking at your face the way Mello does. I have not been able to elicit a slow-blink from them yet. They are both more relaxed cats than Mello, so perhaps they don't need that reassurance (or they don't particularly care!).

It is said that cats may also swallow or lick their nose or lips when they are fearful or stressed, although I have to admit I have never noticed this in real life, hopefully because I have not seen our cats that frightened. When they have been aroused on the rare occasion that a strange cat comes into the house, with all the commotion that that causes, I have been too busy trying to calm the situation rather than looking at their reactions. The three of them also have very different responses to this scenario – Chilli sits back and watches and keeps out of trouble, Mr White makes a lot of noise but doesn't engage, and it takes Mello, the little tabby, to actually make a move and chase the cat out. She will also seem to be more affected afterwards and glance frequently at where the cat appeared from to make sure it's not still there, while the boys simply retire for a nap.

Mello's more 'worried' personality also exhibits when the cats are sometimes given catnip-containing toys by people who kindly think of them when they visit. All of our cats seem to like catnip and react in the usual way, but it is interesting that Mello is more wary of enjoying it if the boys are around. It does seem to lower cats' control and they may get excited or boisterous and play rather roughly. So now I separate them so that she knows she can relax and enjoy it without worrying that she is taking her eye off the others, who may indeed be rather more excitable when catnip is around. She doesn't enjoy even a tiny bit of rough and tumble of the kind the boys are happy to engage in with each other.

Do they know their names?

When Mello and Oreo were kittens, I knew I would have to be able to call them back when it came to that rather nerve-wracking time of letting them go out into the garden for the first time. So, when I fed them or before I did something nice with them, I used a high-pitched call and a certain rhythm, calling 'come on, come on, come on' so they became familiar with it and linked it to something nice. The theory was that even when the big outside was fascinating them, they would still come when called. And, indeed, that did happen – at first, I went out with them and called them in, then gradually let them go out themselves and called them a little while later. The high-pitched 'come on, come on, come on' really worked and they would appear running from all around the garden. I must admit that once they'd become confident going out and I was confident of them being out there, I seldom used the call again. Now I only use it if I haven't seen one for a while and am worrying where they are. It still seems to work, which is quite reassuring.

On a day-to-day basis we use a warm tone of voice with them. I read the research about cats knowing their names – as evidenced by a slight move of the ears or head. In this house we all call the cats lots of different names, Oreo being Mr White and Chilli being Chills or Trouble, and Mello being Mels or Bubs. When my daughter is home she has further names for them, so it probably comes down to tone of voice and the cats actually recognising the different tones and words we use with them individually – they are very bright and observant creatures (they have to be to survive in the wild), so why wouldn't they know we are speaking to them?

When we want our cats to get on our laps we now have a method of patting the place we want them to come to and the cats know they are welcome to get on. We usually wait until we see them come looking to find a lap and then they get the message very quickly. They don't necessarily respond if they are doing something else. They sometimes ignore us, but now we have got to this age, the message gets recognised very easily. It was not always so; in their younger years they would look at us as if we were a bit mad and just carry on, which shows that mutual understanding may take some time to develop and require gentle and simple messages repeated with patience!

They even have their own little foibles – Mr White wants you to put a blanket on your lap first and will wait patiently until you have done that before he gets up, otherwise he will step off, wait and then come back when it's in place. Mello gets confused if you interrupt the ask by putting on a blanket and then the moment may pass and you have to start again. We can also use the pat to encourage them to come to us at other times and out in the garden, meaning they have successfully generalised the understanding to other circumstances.

Enjoy!

I feel rather self-indulgent talking about my cats, but I hope that looking at their individuality helps you to watch and understand your own cats and to enjoy the communication that grows and strengthens over the years. I must admit writing this book has made me look at them more closely and I have enjoyed dissecting their personalities. Even though kittens are irresistible, they are hard work and older cats can be such a joy. Not only have we built up a way to communicate between us, but they completely understand the workings of the household and the things that happen there, fitting in without any fuss.

Cats bring a calm companionship that takes years to achieve. No wonder we miss them so much when they are no longer there. The fact that they are so individual also gives us a great character to remember when they are gone, but it also leaves room for another character to enter our lives, with its own unique personality on which to build our relationship.

Acknowledgements

I WOULD LIKE to thank all of the people from whom I have learned over the years I have been striving to understand cats – the vets, the behaviourists, the researchers, the people working with unowned cats, the enthusiastic owners and those in companies creating medicines and products crafted for cats. As a 'jack of all trades', I have been privileged to benefit from their generous sharing of expertise and experience, and have so enjoyed working with them to do the best we can for cats. Thank you too to all the cats I have owned over the years, and the three who are continuing to train me and feature in this book.

Suggested Reading

CAT DETECTIVE
Vicky Halls
Published by Bantham
ISBN-13: 978-0553816457

CAT CONFIDENTIAL: THE BOOK YOUR CAT WOULD WANT YOU TO READ
Vicky Halls
Published by Bantham
ISBN-13: 978-0553816440

CAT COUNSELLOR: HOW YOUR CAT REALLY RELATES TO YOU
Vicky Halls
Published by Bantham
ISBN-13: 978-0553817621

THE TRAINABLE CAT: HOW TO MAKE LIFE HAPPIER FOR YOU AND YOUR CAT
John Bradshaw and Sarah Ellis
Published by Penguin
ISBN-13: 978-0141979328

THE DOMESTIC CAT 2ED: THE BIOLOGY OF ITS BEHAVIOUR
Dennis C. Turner
Published by Cambridge University Press
ISBN-13: 978-0521636483